Climate Action in Southern Africa

Using climate justice as an analytical tool, this volume examines the role of local mitigation and adaptation actions in Southern Africa in furthering climate-resilient development.

Climate Action in Southern Africa examines the intrinsic connection between local climate actions, climate-resilient development and strides towards a just transition. The theoretical grounding in climate justice allows the authors to analyse whether current climate actions in Africa are truly effective for the poor and marginalised whose lives and livelihoods are impacted by a climate crisis largely not of their making. The authors also question the extent to which pathways to net-zero carbon emissions by 2050 are achievable in Africa and ask whether this can be attained without undermining livelihoods and human development. Overall, this book argues that for any transition to be a just transition, it has to be aligned with the pursuit of sustainable development and climate justice for current and future generations on the African continent.

Drawing out key factors including politics, gender and migration, this volume will be of great interest to students and scholars of climate change, climate justice and African development.

Philani Moyo is Professor of Sociology and Director of the Fort Hare Institute for Social and Economic Research at the University of Fort Hare in South Africa. His primary research interests are in climate justice, adaptation and resilience building, climate-resilient food systems, food justice, food sovereignty and agrarian studies. Some of his recent publications are 'The Political Economy of Zimbabwe's Food Crisis, 2019-2020', *Journal of Asian and African Studies* (2022), DOI: 10.1177/00219096221120923 and 'Contested Compensation: The Politics, Economics and Legal Nuances of Compensating White Former Commercial Farmers in Zimbabwe', *Review of African Political Economy* (2021), DOI: 10.1080/03056244.2021.1990033. His most recent book (co-edited with Akpan, W.) is *Revisiting Environmental and Natural Resource Questions in Sub-Saharan Africa*.

Routledge Advances in Climate Change Research

For more information about this series, please visit: www.routledge.com/Routledge-Advances-in-Climate-Change-Research/book-series/RACCR

Climate Action in Southern Africa

Implications for Climate Justice and Just Transition

Edited by
Philani Moyo

Routledge
Taylor & Francis Group

LONDON AND NEW YORK

First published 2024
by Routledge
4 Park Square, Milton Park, Abingdon, Oxon OX14 4RN

and by Routledge
605 Third Avenue, New York, NY 10158

Routledge is an imprint of the Taylor & Francis Group, an informa business

British Library Cataloguing-in-Publication Data
A catalogue record for this book is available from the British Library

ISBN: 978-1-032-50160-4 (hbk)
ISBN: 978-1-032-50161-1 (pbk)
ISBN: 978-1-003-39712-0 (ebk)

DOI: 10.4324/9781003397120

Typeset in Times New Roman
by codeMantra

Contents

Acknowledgements

All contributors to this book acknowledge financial and administrative support of the University of Fort Hare towards a productive writing retreat that enabled finalisation of chapters. Some of the chapters in this book, where Philani Moyo is the sole or co-author, are products of research funded by the National Research Foundation of South Africa.

Contributors

Adornis D. Nciizah, PhD, is a Specialist Researcher with the Agricultural Research Council of South Africa. He is also a Research Associate with the Department of Agriculture and Animal Health, University of South Africa. His research work focuses on climate-smart agriculture, sustainable bio-energy feedstock production in marginal soils, micronutrient seed priming for micronutrient-deficient soils and sustainable agricultural production in smallholder farming areas. He has been an active team member of various local and international research collaborations funded by the National Research Foundation of South Africa, European Union, Japan International Cooperation Agency and Japan Science and Technology Agency. He has authored numerous scientific papers and reviews on agronomy and soil science, and trained smallholder farmers on climate smart agriculture.

Christopher Dick-Sagoe, PhD, is a lecturer in the Department of Politics and Administrative Studies, University of Botswana. He is a climate action activist and public management expert. His research focuses on local government policy, climate change, livelihoods and youth entrepreneurship. He also has a keen interest on understanding the efficient delivery of decentralised services such as healthcare and education. He accomplishes this by focusing on how available financial resources shape the spending behaviour of decentralised service providers as well as invisible incentives such revenues create for improving the efficiency of decentralised public service delivery.

Stanley O. Ehiane, PhD, is currently a Senior Lecturer in the Department of Politics and Administrative Studies, at the University of Botswana and a Senior Research Associate in the School of Public Management and Governance at the University of Johannesburg in South Africa. He was previously a postdoctoral research fellow at the Fort Hare Institute for Social and Economic Research at the University of Fort Hare in South Africa and also a lecturer and Research Fellow at the University of KwaZulu-Natal, South Africa. His research interests are in the fields of climate change, maritime safety, gender, peace, conflict, terrorism, counterterrorism, cyberterrorism, cybercrime, cybersecurity and hate crime. He has co-published the book *Public Administration: An Introduction,* published the book *Citizenship Education* and also co-edited *Cybercrime and Challenges in South Africa*.

Sipho F. Mamba, PhD, is a lecturer and researcher in the Department of Geography, Environmental Science and Planning at the University of Eswatini in Eswatini. He is also the Lead Researcher in 'Seasonal Livelihood Programming' in the Kingdom of Eswatini and expert in 'Integrated Food Security Phrase Classification and Integrated Content Analysis'. He has researched and published on land and water resources management, food security, urban livelihoods, climate change and variability, and urban and human geography. He is currently the vice president of the Young African Statisticians Association and a member of the Eswatini Focus-Based Financing technical team and Eswatini Vulnerability Assessment Committee core team. He is also a member of the International Society for Urban Health and Food Security Network.

France Maphosa, PhD, is Professor of Sociology in the Department of Sociology at the University of Botswana and a Senior Research Associate in the Institute for the Future of Knowledge of the University of Johannesburg, South Africa. He has previously worked for the University of Zimbabwe, where he was Head of the Department of Sociology for six years, the Biomedical Research and Training Institute of Zimbabwe as a Research Fellow, the Zimbabwe Young Men's Christian Association as Executive Director and the National University of Lesotho as a Senior Lecturer. His research interests include migration and transnationalism, corporate social responsibility, sociology of entrepreneurship, natural resources management, rural and urban livelihoods, labour studies and alternative dispute resolution. His latest publications include *Corporate Citizenship: Business and Society in Botswana*, Palgrave (edited with Langtone Maunganidze in 2021) and *Botswana's Response to the Outbreak of the Coronavirus: Lessons for Disaster Preparedness and Management*, Development Southern Africa (with Israel Rocky Blackie in 2022).

Gracsious Maviza, PhD, is currently affiliated to the International Centre for Tropical Agriculture (CIAT), South Africa as well as the University of Johannesburg in South Africa. She is an ardent qualitative researcher with proven experience in conducting research on migration, gender and inclusion, livelihoods and local development strategies, on which she has several publications. She is currently the Southern African Regional Lead with the CGIAR FOCUS Climate Security Team where she is leading the implementation of the 'Building Systematic Resilience against Climate Variability Extremes' and the 'Fragility, Conflict and Migration' initiatives in Zambia and Mozambique, respectively. She is also co-founder of the Southern Women Academics Network, a network inspired by her experiences and struggles as a female researcher. The chapter where she is a co-author (Chapter 13) was carried out with support from the CGIAR Initiatives on Climate Resilience (ClimBeR), and Fragility, Conflict, and Migration (FCM).

Philani Moyo, DPhil, sole editor of this manuscript, is Professor of Sociology and Director of the Fort Hare Institute for Social and Economic Research at the University of Fort Hare in South Africa. He was previously Head of the

Department of Sociology and Anthropology at the University of Fort Hare, a Teaching Assistant in the School of Politics and International Studies at the University of Leeds (UK) and at the University of Zimbabwe. He has also been a Visiting Professor at the University of Cape Town in South Africa, National University of Science and Technology in Zimbabwe, University of Leeds in the UK and Visiting Associate Professor at Minnesota State University-Mankato, USA. His primary research interests are in climate justice, adaptation and resilience building, climate-resilient food systems, food justice, food sovereignty and agrarian studies. Some of his recent publications are 'The Political Economy of Zimbabwe's Food Crisis, 2019-2020', *Journal of Asian and African Studies*, (2022), DOI: 10.1177/00219096221120923 and 'Contested Compensation: The Politics, Economics and Legal Nuances of Compensating White Former Commercial Farmers in Zimbabwe', *Review of African Political Economy* (2021), DOI: 10.1080/03056244.2021.1990033. His most recent book (co-edited with Akpan, W.) is 'Revisiting Environmental and Natural Resource Questions in Sub-Saharan Africa', ISBN: 1-4438-8651-3, 230 pages, Cambridge Scholars Publishing, Newcastle.

Thokozani P Moyo, PhD, is a sessional lecturer in the Department of Sociology and Anthropology at the University of Fort Hare in South Africa. Her areas of research interest include sustainable rural livelihoods, climate adaptation, gender and women in rural development. She has published on gender and development, and written and presented papers in local and international conferences. As a trained professional teacher, she also has a passion for teaching and engaged scholarship.

Tanaka C. Mugabe, PhD, is currently research manager at the South African International Maritime Institute at Nelson Mandela University in South Africa. Her research interests are in maritime, environment, climate justice and positive youth development. She has written a number of government-commissioned reports and research articles on the maritime sector in South Africa. She is a previous postdoctoral research fellow at the University of Fort Hare and also served on the South African Sociological Association for two consecutive terms as a Council Member.

Treda Mukuhlani, PhD, is a lecturer in the Department of Development Studies, Midlands State University in Zimbabwe. Her main research interests are in climate change, gender, water and security studies. Her expertise is in water governance, gender and disaster management. In addition to the above research areas, she has also authored and co-authored articles and book chapters on peace and security, constitutionalism and conflict management.

Elinah Nciizah, PhD, is a lecturer in the Department of Development Studies, Midlands State University in Zimbabwe and is also affiliated to the School of Public Management, Governance and Public Policy, University of Johannesburg, South Africa. She has lectured for over 12 years and has vast experience on developmental issues that include climate change, human rights, poverty, inequality, rural policies, discrimination against women and marginalisation.

She researches and publishes on rural development covering rural livelihoods, climate change, food security, disaster risk management, water issues and agriculture. She has been involved in Community Needs Assessments and Rural Industrialisation Initiatives in rural areas of Zimbabwe.

Tendai Nciizah, PhD, is a lecturer in the Department of Development Studies at Great Zimbabwe University in Zimbabwe and also affiliated to the Barefoot Education for Africa Trust, Zimbabwe. She has research experience on climate change adaptation, rural livelihoods, environmental management, land degradation, small grains production and food security. She has published works on climate change, food security, land issues and environmental degradation. She has previously worked with various non-governmental organisations in Zimbabwe on climate change issues, gender and equity, and land and food security.

Thabo Ndlovu, PhD, is a Senior Lecturer at the Institute of Development Studies, National University of Science and Technology (NUST) of Zimbabwe and a Disaster Risk Management and Humanitarian Specialist. His research interests are in climate smart agriculture, climate mitigation, adaptation, resilience building and asset creation. He has published on sustainable agriculture, irrigation farming, disaster risk reduction and cash transfer modalities. Since 2018 to date, he is the Project Coordinator of a joint research project of the Institute for Development Studies-NUST and United Nations World Food Programme, Zimbabwe office. Further, he is a member of the Africa Science Technology Advisory Group on Disaster Risk Reduction under the African Union Commission. He was previously a research team member in projects that include 'Strengthening of Research Collaboration on Disaster Risk Reduction in Southern Africa' funded by the World Bank/SADC and participated in development of World Food Programme Global Guidance Manuals to inform resilience planning in rural and urban settings. He is also a Steering Group member of the Climate Mobility Africa Research Network.

Divane Nzima, PhD, is affiliated to the University of the Free State as a lecturer and subject head in the Department of Sociology at the Qwaqwa campus in South Africa. His research and teaching interests cut across various disciplines such as Sociology, Demography and Development Studies. He has published research on the intersection of migration and development, family dynamics and parenting, wellbeing of the elderly in resource-constrained countries as well as artisanal mining and agricultural value chains.

Thapelo Ramalefane, PhD, is a post-doctoral research fellow at the Fort Hare Institute for Social and Economic Research, University of Fort Hare in South Africa. His research interests are in climate finance and resilience, livelihoods and positive youth development. His research work has focused on rural communities in Maseru (Lesotho) and in the Eastern Cape of South Africa. He has completed grant-funded research projects for organisations that include Red Cross in Lesotho and the Department of Basic Education, Eastern Cape Province in South Africa.

Part I
Overview

1 The Climate Crisis and Climate Injustice in Southern Africa

Philani Moyo

Introduction

The climate crisis is not just a human development challenge. It "is an existential threat to humanity and the planet and a cruel engine of myriad forms of injustice, disruption and destruction" (Borras *et al.* 2021: 1). Despite numerous global climate action initiatives, United Nations Framework Convention on Climate Change (UNFCCC) declarations and plans of action, the Paris Agreement on Climate Change and national-level climate policies and programmes, global warming continues without any decisive tangible solution in sight. "Instead, we have continued emitting pollutants and intensively using fossil fuels and, as a result, have been recording the hottest years on the planet" (Satgar 2018: 1) as "global emissions are at their highest level in history – and rising" (United Nations 2022: 1). This is unsustainable, and this human-induced climate crisis cannot be allowed to continue unfettered. There is thus an urgent need to halt this climate system destruction through decisive climate action as aptly articulated by the United Nations (UN) Secretary General António Guterres when he said:

> I am here to appeal to all parties to rise to this moment and to the greatest challenge that humanity is facing. The world is watching and has a simple message to all of us: stand and deliver. Deliver the kind of meaningful climate action that people and planet so desperately need....Climate impacts are decimating economies and societies – and growing. We know what we need to do and we have the tools and resources to get it done.
>
> (Guterres speech at the Conference of the Parties (COP27) meeting in Egypt, United Nations 2022: 1)

These remarks remind us that the anthropogenic climate crisis, with its global implications, is impacting interconnected aspects of our material existential conditions. Agriculture, food security, water security, human health, physical infrastructure, human settlements, energy systems, ecosystems and biodiversity, etc. are all being decimated by the climate crisis at a global scale. The magnitude of its destruction warrants characterization of the climate crisis as an emergency due to the dire consequences it is inflicting on humanity.

DOI: 10.4324/9781003397120-2

While the impacts of the emergency are global, they are uneven and unpredictable – uneven in terms of impact inter- and intra-country (or community) and unpredictable geospatially (Borras *et al.* 2021). As a geophysical phenomenon, Newell (2022: 915) further explains, "climate change exacerbates existing social inequalities and vulnerabilities around access to land and water, for example, as well as livelihood security through disruption to ecosystems and infrastructures and via extreme weather events". This means the environmental, social, economic, health and political impacts produced by the anthropogenic climate emergency are experienced at various geospatial locations in different magnitudes depending on levels of vulnerability, inequity and inequality. Its most devastating impacts are disproportionately visible and most vicious in the developing world (Global South), more so on the vulnerable, poor and marginalized. Their suffering, according to Mwenda and Bond (2020), signifies 'climate apartheid' because these people, who are now unevenly and excessively at the 'coal-face' of the crisis, contributed very little to its genesis. Instead, the crisis is primarily a product of Western systemic capitalist accumulation and, according to Borras *et al.* (2021: 2), "industrialism associated with state socialism". This entanglement of capitalism and the climate crisis is historical, emerging from the time of the Industrial Revolution in now developed countries (Global North) which changed the trajectory of greenhouse gas (GHG) emissions, accelerating during epochs of state-led socialism to full throttle under contemporary neoliberal globalization. The evolution of industrial production processes powered by fossil fuels and multifaceted energy uses behind economic growth in the North thus altered the global climate system through unprecedented GHG emissions. This Anthropocene, engineered by contemporary capitalism, consequently broke "a 10 000-year pattern of relatively stable climate known as the Holocene" (Satgar 2018: 3). This relationship between capitalism, industrialization and unprecedented GHG emissions by the North is the basis of the argument that it carries the historical burden of the genesis and acceleration of the climate crisis. It is also this historical fact that informs our view that it is incumbent on the North to be mainly responsible for supporting mitigation, adaptation and resilience actions in the South within the context of "common but differentiated responsibilities and respective capabilities, in the light of different national circumstances" (United Nations 2015: 3) as enunciated in the Paris Agreement.

Within this political economy of the climate crisis, developing countries of the South, environmental and climate justice non-governmental organizations (mainly from the South) and grassroots people's movements are at the forefront of advocacy and radical push for climate justice for people of the South. Although there are multiple climate justice narratives in this space, for our purposes, we argue that climate justice is

> used to account for and contest how climate change is having the most severe effects on those with the least responsibility for causing it, and who, at the same time, are often excluded from decision-making processes regarding

responses to the problem, whether with regard to climate mitigation or adaptation.

(Newell 2022: 916)

Therefore, "climate justice is the recognition that the historical responsibility for the vast majority of greenhouse gas emissions lies with the industrialized countries of the Global North" (Petermann 2009: 135–136). This means the starting point of our climate justice narrative is the observation, and acknowledgement, "that inequalities and injustice are at the root of the causes and impacts of climate change" (Borras *et al.* 2021: 12). Those least responsible for it now suffer from a range of deeper vulnerabilities, risks, stressors, inequities and inequalities. At the centre of this injustice is the structural marginalization and exclusion of the poor from polity, mainstream socio-economic life as well as the enduring relationship between capitalism and destruction of the climate system. For these reasons, the neutral apolitical view taken by some neoliberals about what constitutes climate justice does not resonate with our narrative. For example, we disagree with the argument by Preston *et al.* (2014: 3) that climate justice is merely about individual and collective climate action that simply takes account of different vulnerabilities, resources and capabilities without delegating primary responsibility for these actions. This narrative is ahistorical and apolitical. Its assumption of equality in individual and collective climate action is oblivious of the marginalization, inequity and injustice emanating from the climate crisis.

There are various, mostly overlapping, climate justice approaches. Within the context of our working conceptualization of climate justice, three approaches, namely, the rights-based approach, sustainable development approach and the historical responsibility perspective, frame our analysis. The liberal rights approach underscores climate change consequences that "threaten civil and political rights and economic, social, and cultural rights, including rights to life, access to safe food and water, health, security, shelter, and culture" (Levy *et al.* 2015: 310). The reasoning is that deprivation or violation of these rights, because of the climate emergency, is a grave injustice. As a corrective measure, rights-based approaches argue for mitigation and adaptation compensation for those most affected. This reparation position echoes advocates of the historical responsibility approach. Their "more radical approach to justice emphasises the rights of those already structurally marginalized, including the poor and future generations" (Tschakert and Machado 2012 cited in Borras *et al.* 2021: 12). Their argument is that the injustice endured by the poor is a by-product of inequitable distribution of climate vulnerability, risks and impacts that are compounding their preexisting conditions. Climate impacts' inequity is thus an additional burden to the broader challenges and reality of their everyday life. In an effort to address this intergenerational injustice, historical responsibility advocates ask: who and how was this injustice constructed? What should be done and by whom?

Within the foregoing climate justice discourse, it is common (as we hereby do) to deploy procedural and distributive justice as analytical tools. Distributive justice is concerned with how resources, benefits, burdens and outcomes are distributed across people, within countries, between generations and whether a distribution is morally acceptable (Ikeme 2003; Jafino *et al.* 2021). This distribution "can be based on an assessment of responsibilities, rights or needs, ability, capacity and economic efficiency" (Preston *et al.* 2014: 14). Analogously, procedural justice calls for diverse people's active participation, inclusion, fairness and transparency in mitigation programming and decision-making processes (Preston *et al.* 2014). These participatory justice concerns extend to local-level adaptation as grassroots movements agitate for inclusion in adaptation decisions and processes of programme planning, implementation and resilience-building post-climate disasters (Schlosberg and Collins 2014). These corresponding arguments of procedural and distributive justice are thus ideal in exploring the different dimensions of climate injustice in Southern Africa. They provide a platform for us to ask: who is involved and excluded in climate action decision-making (procedure)? Who is benefiting from and who is losing due to the climate emergency (distribution)? What corrective action must be taken to compensate those affected the most (reparations)?

Climate Justice

Climate justice discourse is largely informed by ideas and concepts of the environmental justice movement. Over time, the environmental justice movement underwent discursive shifts in its conceptualization culminating "in the very first principle of environmental justice affirming the sacredness of Mother Earth, ecological unity, and the interdependence of all species" (Schlosberg and Collins 2014: 361). Embedded in this principle, which dismissed the traditional narrow conception of environment as wilderness disconnected from human life (Wright 2011), are the environmentalism of everyday life, "relationship between human beings and nonhuman nature…relationship between cultural practices, sovereignty rights, and lives immersed in diverse and threatened ecosystems" (Schlosberg and Collins 2014: 361). These central elements of environmental justice alongside ideas about the distribution of environment-induced vulnerability and the extent to which environmental risks threaten humanity are some of the foundational ideas of environmental justice that 'gave birth' to climate justice. Just like in environmental justice thinking, social justice demands are central to climate justice.

There is a broad and pluralistic remit of climate justice approaches. For our purposes, we borrow ideas from overlapping conceptualizations that include rights-based approaches, sustainable development approach and the historical responsibility perspective. Rights-based approaches to climate justice, which are an element of procedural justice, "give primacy to human rights, duties, obligations, responsibilities and fairness of procedures" (Preston *et al.* 2014: 13). Their central argument is that impacts of the climate emergency violate various basic human rights, of the poor and marginalized, as embodied in international and national legal frameworks. For example, protections offered by Article 25 of the Universal Declaration of Human Rights and Article 11 of International Covenant on Economic, Social and Cultural Rights are infringed by climate impacts. These include the right to food, medical care, good health, housing, social services and social security. Contrary to Moellendorf's (2012: 134) view that these are demanding and controversial universal rights, we argue that these are fundamental basic human rights that ought to be respected, defended and preserved at all times. This stance echoes Caney's (2010) argument that basic rights that include the right to life, human security and right to preserve and sustain subsistence means, all of which are violated by climate change, must be defended and preserved. In their unjust manner, the calamities of the climate emergency are thus disrupting or denying the poor these rights, thereby providing a normative justification for advancing and advocating for climate justice in order to recognize and respect these rights.

As a corrective intervention, rights-based approaches advocate for procedural justice as one pathway towards climate justice. This dimension of procedural justice is, among other issues, concerned with "who is involved in decision-making and how are past wrongs addressed" (Borras *et al.* 2021: 12) through mitigation and adaptation actions. Also central to procedural justice are demands for active inclusion and participation by vulnerable communities in climate action decision-making and programming (Schlosberg and Collins 2014: 361) as well as "fairness in and legitimacy of these planning and decision-making processes" (Jafino *et al.* 2021: 2). This thinking emerges from a realization that the historical exclusion of poor and vulnerable communities from participation and decision-making has previously enabled inequity and inequality on mitigation and adaptation matters that directly impact their lives. For these reasons, procedural justice is thus vital as it ensures local communities are actively included in conversations from the onset, collectively make decisions and deepen local empowerment. Similarly, it enhances "self-determination, sovereignty, human rights…" (Schlosberg and Collins 2014: 361) as part of social justice and a conduit to climate justice.

The rights-based approach overlaps with the sustainable development perspective that emphasizes prioritization of the poor's basic needs, limitations of the environment in meeting these needs and developmental rights. It is widely accepted that sustainable development refers to "development that meets the needs of the present without compromising the ability of future generations to meet their own needs" (Brundtland 1987: 8). Although this definition has been criticized heavily (see Mitlin 2016), it remains the best starting point in understanding the intersection between socio-economic development, sustainability and the climate system. Contrary to those who argue that economic growth and sustainable development are incompatible since "continued economic growth will lead (sooner or later) to environmental degradation and a lack of development opportunities" (Mitlin 2016: 111), the reality is that sustainable economic development, in the North and South, can be, and must be, pursued in a physically, socially and environmentally sustainable manner. This includes sustainable use of resources, distribution of their costs and benefits, and concern for socio-equity within and between generations (Brundtland 1987; Mitlin 2016). This idea of social equity between generations echoes distributive justice thinking which is also concerned with how development goods, natural resources, services and benefits are assigned intra and inter-country (Lamont 2016). In a similar vein, development rights advocates argue that "all people and nations should have a right to develop out of poverty" (Schlosberg and Collins 2014: 365) in a manner that respects, safeguards and sustains environmental, climatic and human rights of current and future generations. That is to say, as part of intergenerational justice, it is the right and duty of the current generation to utilize natural resources for their own human development but in a sustainable way that will not deprive future generations of similar benefits.

The nuance of this intergenerational justice is further articulated through a distributive justice lens. Distributive justice identifies a number of climate injustices related to "who gets which benefits and who suffers what costs/risks" (Borras *et al.* 2021: 12), how climate action "resources, benefits and burdens are allocated between or within countries or between generations" (Ikeme 2003 cited in Preston *et al.* 2014: 14) and the "distributional outcomes of mitigation and adaptation policies" and actions (Jafino *et al.* 2021: 2) at national and local levels. This focus allows an assessment and identification of how the climate crisis is creating or deepening ecosystems and biodiversity destruction, undermining human capabilities, their basic needs and devastation of environmental resource-linked livelihood activities of communities in the South. Further, these lenses allow an understanding and provide an avenue for beginning to address unsustainable environmental actions, poverty, inequity, inequality and injustices induced by the anthropogenic climate emergency.

Further, the sustainable development perspectives' concern about how resources are used for intergenerational benefit is the driving force behind the repositioning of the nonhuman realm (natural world) in the discourse of what constitutes climate injustice and who caused it. This reconsideration has broadened the dialogue beyond inequity of impacts on human beings. As a result, many critical reflections now "also address the ecological damage done to surrounding ecosystems that have led to greater vulnerabilities for both human communities and the nonhuman environment" (Ross and Zepeda 2011 cited in Schlosberg and Collins 2014: 363). The placement of the nonhuman environment at the centre of the debate emerged out of recognition "that the environment and climate system are not simply symptoms of existing injustice, but instead the necessary conditions for the achievement of social justice" (Schlosberg and Collins 2014: 363). This means for climate justice to be realized, there must be structural and behavioural changes in the manner in which the environment is used in the value chain of meeting human needs. Realization of sustainable environmental use, for the benefit of current and future generations, is thus a pathway towards reducing destruction of the climate system.

Some elements of the sustainable development perspective overlap with the historical responsibility approach that has specific demands for achieving climate justice. With overwhelming support from developing countries of the South (including Southern Africa), it emerges from and is informed by the 'polluter-pays-principle' (PPP). In basic terms, the PPP is an economic, ethical and legal principle that ties "responsibility for adaptation and mitigation and for generating reliable funding for the purpose" to countries of the North that primarily produced and continue to drive the climate crisis (Khan 2015: 638). The PPP stance is thus an attempt to address a global commons problem (the climate emergency) through an equitable process (or system) that fairly distributes responsibility. The idea is that the North must take responsibility through concrete climate actions for the mutual benefit of both North and South. On the surface, this seems like an impartial argument. However, it is contested, faces resistance from the North and is loaded with contradictions. One of its glaring

> contradictions is that while it rests on neoliberal market principles, the UN Framework Convention on Climate Change (UNFCCC) did not include the PPP as its provision though the principle of 'common but differentiated responsibility based on respective capabilities' (Article 3.1).
>
> (Khan 2015: 638)

Instead, the UNFCCC's Article 3.1 implicitly assumes this. Nevertheless, the historical responsibility approach draws on dimensions of the PPP and the UNFCCC's Article 3.1. It further invokes the Paris Agreement's Article 2.2 which urges parties to the Agreement to implement it to "reflect equity and the principle of common but differentiated responsibilities and respective capabilities, in the light of different national circumstances" (United Nations 2015: 3). Within this premise, the central argument of historical responsibility is that actions of the North brought us to the current climate emergency, and hence, they should shoulder primary responsibility for the consequences of their actions. This allocation of responsibility is informed by their historic excessive emission of GHG emissions. For that, they "should pay the costs caused by these past transgressions" (Schlosberg and Collins 2014: 365) as a form of compensation for the loss and damage inflicted on the vulnerable, marginalized and poor of the South. This payment "reflects the most fundamental principles of justice and responsibility" (Khan 2015: 638) and is not an act of benevolent charity, from the North to the South but a reparation for the environmental and climate system damage they caused.

Unsurprisingly, there are organized political and business interests in the North opposed to this historical responsibility based on superfluous and misplaced arguments. We reject their arguments, especially those aligned to Moellendorf's (2012: 135) view that assigning historical responsibility is problematic for a number of reasons:

> first, because of the long life of CO_2 molecules in the atmosphere, much of the damage-causing stock of atmospheric CO_2 was produced by people who are now long dead, and from them no costs can be recouped. Second, among the still-living, many fail to meet the knowledge condition for some of their early emissions.

His, and those of similar persuasions, defence of the biggest CO_2 emitters is ahistorical and lacks factual basis. The fact, and reality, is that as part of ensuring procedural and distributive justice, poor people affected by climate impacts have a right to "get redress" (Preston *et al.* 2014: 14) from those who created the crisis and have the financial and technical resources to pay. Ultimately, this is not just a moral course of corrective action to take but also the ethical thing to do. It is also a settlement of the 'climate debt' they owe the South. This climate debt is a redistributive and compensation instrument for "violation of communal rights and territories" (Warlenius 2018: 132). It is intertwined with historical responsibility in the demand for "justice for those communities that are on the front lines of the climate crisis, but who have contributed least to it" (Newell 2022: 915). The North thus owes this climate debt to the South "as a result of their greater historical contribution to human-induced climate change" and because "the resulting need to reduce emissions globally now constrains the ability of poorer countries to develop" (Pickering and Barry 2012: 667). This debt should be repaid through a mechanism that places the burden of mitigation and adaptation costs on the North (Warlenius 2018: 132) as well as by ensuring that developed countries rapidly reduce their emissions and provide "finance to help developing countries adopt low-emissions technologies" (Pickering and Barry 2012: 667–668). However, there are many senior bureaucrats, politicians and business interests in the North who oppose the idea and rationale for climate debt. They dismiss legitimate claims for climate debt offhand, "principally by means of two arguments: that the idea of a climate debt makes no sense or that employing climate debt as a frame for understanding climate related responsibilities will lead to morally objectionable conclusions" (Pickering and Barry 2012: 668). Likewise,

> a great many Big Green Groups in the United States consider the idea of climate debt to be politically toxic, since, unlike the standard 'energy security' and green jobs arguments that present climate action as a race that rich countries can win, it requires emphasizing the importance of international cooperation and solidarity.
>
> (Klein 2014 cited in Bond 2022: 15–16)

This resistance to climate debt by some in the North is a manifestation of "unjust power relations between North and South" (Warlenius 2018: 132) in the climate justice space. Powerful interests in the North simply have no appetite or climate diplomacy desire to subject themselves to matters of equity and equality spearheaded by the South. Relatedly, the profit accumulation interests of economic forces that fund the work of Northern anti-climate debt lobbyists and some elite Big Green Groups explain this opposition to climate debt. Their resistance ignores the fact that the responsibility of climate debt is the burden of those primarily culpable for past climate transgressions. Their violation of the environment, communal ecosystems, biodiversity and climate system therefore gives rise to this call and struggle for compensatory climate debt.

Climate Injustice in Southern Africa

Given the continuing decimation of the climate system and its human development costs, it is befitting to characterize the climate crisis in Southern Africa as an emergency due to the dire consequences it is inflicting on humanity, more so on the vulnerable, poor and marginalized. While cautious not to caricature the climate emergency nor paint a picture of catastrophism, we are unapologetic in arguing that the current globalized system of neoliberal capitalism (dominated/led by the North) with its insatiable fossil fuel-powered production and consumption is driving Southern Africa to the precipice unless production and consumption is radically transformed. This systemic transformation, which should begin in the North, must be driven by and aim for climate justice for people of the region.

This struggle for climate justice in Southern Africa is provoked by glaring injustices driven by the climate emergency in the region. Firstly, drawing on empirical cases from across the region, it is clear that the already vulnerable, poor and marginalized leading precarious livelihoods due to the failed Western-initiated neocolonial project of neoliberal economic structural adjustment programmes (ESAP) are disproportionately suffering the most. Preexisting injustices in the form of poverty, unemployment and inequality from the free market, privatization and deregulation of ESAP disaster years are being deepened by the climate emergency. Further, these communities are currently underserved in the sense that their governments are failing to provide adequate adaptation and resilience support in the form of information, capacity building and material resources thus perpetuating injustices. Secondly, and relatedly, failed capitalist-oriented macro-economic policies and misgovernance by postcolonial governments in the region are exacerbating poverty and economic hardship with climate impacts making existential conditions direr. This is happening in a socio-political context where the grassroots have limited (or no) voice and hardly participate in top-level decision-making. This is a form of procedural injustice since voices of the poor and marginalized communities are not adequately included in political and climate action decision-making. Thirdly, as alluded above, the people of Southern Africa are less responsible for GHG emissions that continue to devastate the climate as they are emitting a tiny fraction compared to the global average. Instead, historical and current GHG emission statistics demonstrate that Northern countries (and of late China and India) primarily emit the most GHG and are thus the drivers of the climate emergency. Finally, mitigation and adaptation policies in some Southern African countries, as embedded in their climate action policies and strategies, are a potential threat to jobs, especially those in the mining-manufacturing industrial complex powered by fossil fuels. While the anticipated climate benefits of a transition to renewable energy are not in dispute, the cost of this transition on the working class, their jobs and livelihoods is a matter of potential injustice. Given that there is currently no evidence of the transition opening lucrative new frontiers of job creation in the form of 'green jobs' or local development of new renewable technologies, the probability of an inequitable transition is high. This calls for the structure and programming of just transition in the region to be cognizant of and avoid inequitable unintended impacts.

Further, in framing, questioning and responding to this injustice, we use overlapping climate justice principles (flowing from the foregoing climate justice concepts) to structure our analysis herein. Although not exhaustive, we argue that climate justice must be based on radical reduction of the use of fossil fuels accompanied by a gradual just transition to renewable energy in an equitable manner, protecting poor and vulnerable communities from unjust and inequitable climate impacts, leveraging just transition towards human development of the least well-off, ensuring active community participation in climate action decision-making and programming (Schlosberg and Collins 2014). This characterisation of climate justice is opposed to neoliberal growth economics and industrialization strategies that have destroyed the climate system. It also articulates sustainable human–nature interactions to mitigate the Anthropocene, challenges the patriarchal and exploitative nature of neoliberal capitalism and advances the need for sustainable, just and fair climate action. Further, we argue that for climate justice to be realized in Southern Africa, it is imperative not to acquiesce with apolitical analysis of climate impacts, mitigation and adaptation. Those who view and analyse the climate emergency through apolitical lens have a parochial understanding of how the politics of production, accumulation and consumption are the epicentre of creating, driving and perpetuating climate injustice in the region. We proceed to state that an equitable climate justice pathway ought to entail the North supporting climate actions of poor communities, advance the struggles of agrarian communities, urban poor and working class towards sustainable resilience building. This reparation mechanism should be part of coordinated efforts towards intergenerational justice. In our view, such a climate justice pathway has potential to compensate for preexisting, existing and impending injustice linked to climate-induced vulnerability and address injustice in the dispersal of climate impacts, their costs (human or otherwise) and existential conditions of the poor.

Our climate justice narrative is informed by empirical evidence and secondary analysis from the frontiers of climate injustice and climate politics in Southern Africa. The second chapter, by all contributors in this manuscript, is a policy discussion that unpacks and analyses selected climate laws, policies, strategies and action plans of Botswana, Eswatini, Lesotho, South Africa and Zimbabwe with a view to understanding the extent to which they are pathways towards social and climate justice. This analysis paints a broad picture of the state of climate injustice as the climate crisis continues to unfold. In Chapter 3, Thabo Ndlovu and Philani Moyo draw insights from the sustainable development perspective as they examine how climate-induced droughts are contributing to diminishing communal rangelands thus constraining communal livestock farmers' adaptive capacities in Umzingwane District in Zimbabwe. They argue that due to the pressing climate emergency, sustainable rangeland management is not practised leading to chaotic local grazing strategies and schedules. While there are by-laws and traditional norms that should ideally shape local practices, regulate communal natural resources and biodiversity preservation, these are not in sync with climate justice pathways since they are not prioritizing local equity and sustainable livelihoods of marginalized livestock farmers. The debate on climate injustice in Zimbabwe is extended by Treda Mukuhlani in Chapter 4. Through her empirical case study of water insecurity in Bulawayo, she argues that at a macro level, the government of Zimbabwe is not pro-climate justice on climate adaptation and resilience matters. She then deploys the rights-based approach in examining the intersectional and complex ways that climate vulnerability, risk and impacts are affecting the urban poor's access to water. She adds that Bulawayo's poor also perennially suffer from water insecurity due to neoliberal inclined water governance systems and politicization of bulk raw water interventions designed to solve the crisis. Seen from a liberal rights prism, she argues, this deprivation of access to water is a violation of the human, economic and social rights of poor residents of Bulawayo.

The fallacy of neoliberal prescripts in building community-level resilience is further exposed through exploring climate finance public–private partnerships in Lesotho. In Chapter 5, Thapelo Ramalefane and Philani Moyo examine the climate finance architecture in Lesotho, with a specific focus on its modalities at a local community level, bureaucratic politics and inefficiencies of public–private partnerships in the smallholder citrus farming sector. Using sustainable development perceptions, they argue that the top-down decision-making model of public–private partnerships is not an antidote to the challenges faced by smallholder citrus farmers in Lesotho. This is because neoliberal-informed public–private partnerships are structurally and institutionally chaotic with poor coordination. The limited and erratic climate finance compounds inefficiencies of the public–private partnership hence its failure to build smallholder citrus farmers' resilience. Using the historical responsibility lens, they strongly criticize loans advanced by the World Bank under this public–private partnership. They argue that World Bank loans cannot be a panacea to the climate emergency because, rather than solve the problem, they are further deepening Lesotho's debt crisis. The World Bank, through historical financing of environmentally unfriendly growth economics in the North, actively begot genesis of the climate crisis and cannot be expected to be a neutral arbiter in seeking equitable adaptation and resilience solutions in Lesotho.

From Lesotho, the discussion moves to Eswatini in Chapter 6. Here, Sipho Felix Mamba and Thabo Ndlovu advance just transition and sustainability debates with a view to understanding their implications on poverty reduction, inequality and jobs. While framing their examination in historical responsibility concepts, they argue for a South perspective of the just transition, one which considers the contextual realities and precariousness of their economies characterized by high poverty rates, unemployment and persistent inequalities. They argue that even though socioeconomic issues such as poverty reduction, employment creation and inequality reduction are embedded in the just transition policy framework of Eswatini, it remains unclear whether the country has economic wherewithal and socio-political desire to abandon its current development trajectory in favour of a North model of just transition. Within the just transition pathway, a detour to Zimbabwe is done by Tendai Nciizah, Elinah Nciizah and Adornis Nciizah in Chapter 7. Their secondary review of climate actions and just transition attempts in Zimbabwe unravels contradictions of mitigation and adaptation in cases where some privileged people (and profit-driven institutions) reap socio-economic benefits, while the poor's livelihoods are catastrophically destroyed. As they argue for a just transition that builds local economic and political power with the aim of shifting the country from an extractive to a regenerative economy, they note that this transition will result in closure of industries that are physically harmful to workers, their health, community health, ecosystems and biodiversity while simultaneously providing just pathways for workers to move into new diversified 'green' livelihoods. In doing this, they use insights of the capabilities approach and rights-based perspective both of which aim at bringing social and political voice, recognition and agency of the poor and vulnerable.

The rights of the poor and vulnerable, in so far as they relate to climate governance, are broadened in Chapter 8 by Philani Moyo. Focusing on climate governance, inaction and injustice in Buffalo City, South Africa, he argues that at the centre of the city's failed climate governance system is the fact that its climate conscious integrated development plans and climate action strategies remain 'on paper' without noteworthy practical measures or steps being taken towards full implementation. This failure emanates from the intersection of polity and politics that have a direct bearing on how city climate action priorities are decided. Where polity and political dynamics favour pursuance and delivery of traditional basic services and infrastructure in order to retain power at the city hall, these considerations supersede the need for climate actions in response to the climate emergency. Although this might appear, at face value, to be a noble decision taken in the context of competing interests between local development needs and climate action, it is in fact symptomatic of the city's failure to build an integrated development ecosystem that meets local human development needs while simultaneously driving a mitigation, adaptation and resilience agenda towards climate justice. Further, despite mainstreaming climate action in its local development plans and strategies, the city is failing to adequately budget and allocate climate finance. Consequently, climate programming remains fragmented as the city is failing to effectively implement *ex ante* mitigation and *ex post* adaptation, especially in poor and vulnerable townships, thus deepening climate injustice.

The scourge of climate injustice in South Africa is further explored in Chapter 9 by Philani Moyo. In this empirical analysis, he examines the effectiveness of adaptation strategies implemented by smallholder farmers as they seek to build resilience against climate change impacts in the Eastern Cape, South Africa. He argues that while the agency and ingenuity of smallholder farmers is not in question, their adaptation is constrained by underlying vulnerability drivers that include limited physical, human, financial and social capital with direct implications on social and climate justice. With variations across households, access to different forms of capital determines the character and extent of challenges endured. This is personified by how limited financial capital is partly responsible for farmers' inability to enhance adaptation through irrigated horticulture, procurement of livestock with drought tolerant genetics and failure to buy silage equipment for fodder production. Similarly, human capital constraints partly explain why many cannot access relevant climate variability data that consequentially leads to arbitrary decision-making in adaptation crop choices and farming practices. Relatedly, social capital constraints mean resilience-building information shared within their social networks is of limited scope and utility, thus constraining effectiveness of their adaptation strategies.

In Chapter 10, Tanaka Mugabe provides a wealth of insights into understanding the theoretical premises of just adaptation and how it relates to climate justice. She also deploys systems thinking to understand the nexus between social, economic and environmental factors that impact youth and climate justice in rural South Africa. She argues that numerous possibilities exist for rural youth to use their agency, skills and knowledge to pursue just adaptation even as the climate crisis continues to rage. The interplay of human development and climate injustice is explored by Stanley Ehiane and Christopher Dick-Sagoe in Chapter 11. In their exposition of climate change and injustice in the context of Botswana's national adaptation actions, they argue that the international community should take the lead in ensuring that climate injustices are alleviated. For this to happen, the North must work towards climate equity through providing climate finance to less developed countries such as Botswana. At a local level, the participation of marginalized groups that include women, youth and civil society organizations (procedural justice) alongside government bureaucrats in climate actions is advanced as a step towards climate justice. In unison, Acts of Parliament that safeguard human rights of the poor and vulnerable in the context of a changing climate should be enacted as a pathway towards climate justice. They conclude that for all this to be realized, political will in Botswana's polity and political landscape will be a deciding factor.

This book concludes with a reflection on the climate change–migration nexus in the region. In the penultimate Chapter 12, France Maphosa analyses climate mobility and social justice in Matabeleland, Zimbabwe. He argues that although migration scholarship generally agrees that climate impacts contribute to migration, establishing a linear causal link between climate factors and migration has not been possible. This is because the decision to migrate or not in the face of the climate emergency is also influenced by other factors which revolve around issues of social injustice. As a result, migration in the context of the climate emergency does not constitute one single act but is a continuum in a wide range of mobilities, including immobilities. In this context, climate mobility lens therefore provides a useful framework for understanding the dynamics between human mobility and climate as well as the connectedness of environmental and socio-economic issues. A gender dimension is brought into the climate mobility debate by Divane Nzima and Gracsious Maviza in the last chapter, which is Chapter 13. Their chapter embeds a gender lens to assess how the climate emergency influences gendered migration patterns in Southern Africa. It notes that as rural agro-based livelihoods are destroyed by frequent droughts, high temperatures and water insecurity, women who have traditionally remained behind when men migrated have joined the exodus to build resilience and ensure social equity. Some of the women are engaged in cross-border trading, while some seek seasonal jobs in commercial farms in neighbouring countries. Since rural women are disproportionately affected by the climate emergency, they have chosen migration as a strategy to self-insure against livelihood risks. Therefore, women now form a significant component of the increasing migration stocks, leading to growing feminization of migration.

References

Bond, P. (2022) 'Loss and Damage Calculations, Polluter Liabilities and Demands for Climate Debt and Reparations', Unpublished Workshop Paper, Presented at the Friedrich Ebert Stiftung Foundation Workshop on the 'African Union-European Union Green Deal', 9–11 June, Harare: Cresta Lodge.

Borras, Jr. S.M., Scoones, I., Baviskar, A., Edelman, M., Peluso, N.L. and Wolford, W. (2021) 'Climate Change and Agrarian Struggles: An Invitation to Contribute to a JPS Forum', *The Journal of Peasant Studies*, Vol 49 (1), 1–28, https://doi:10.1080/03066150.2021.1956473.

Brundtland, G.H. (1987) *The United Nations World Commission on Environment and Development: Our Common Future*, Oxford: Oxford University Press.

Caney, S. (2010) 'Climate Change, Human Rights and Moral Thresholds', in S. Humphreys (ed.), *Human Rights and Climate Change*, Cambridge: Cambridge University Press, pp. 69–90.

Ikeme, J. (2003) 'Equity, Environmental Justice and Sustainability: Incomplete Approaches in Climate Change Politics', *Global Environmental Change*, Vol 13 (3), 195–206.

Jafino, B.A., Kwakkel, J.H. and Taebi, B. (2021) 'Enabling Assessment of Distributive Justice Through Models for Climate Change Planning: A Review of Recent Advances and a Research Agenda', *WIREs Climate Change*, Vol 12 (4), https://doi.org/10.1002/wcc.721.

Khan, M.R. (2015) 'Polluter-Pays-Principle: The Cardinal Instrument for Addressing Climate Change', *Laws*, Vol 4, 638–653, https://doi:10.3390/laws4030638.

Lamont, J. (2016) *Distributive Justice*, London: Routledge.

Levy, B.S., Patz, J.A., Sherborn, M.A. and Madison, W.I. (2015) 'Climate Change, Human Rights, and Social Justice', *Annals of Global Health*, Vol 81 (3), http://doi.org/10.1016/j.aogh.2015.08.008.

Mitlin, D. (2016) 'Sustainable Development: A Guide to the Literature', *Environment and Development*, Vol 4 (1), 111–124.

Moellendorf, D. (2012) 'Climate Change and Global Justice', *WIREs Climate Change*, Vol 3 (2), 131–143, https://doi.org/10.1002/wcc.158.

Mwenda, M. and Bond, P. (2020) 'African Climate Justice: Articulations and Activism', in B. Tokar and T. Gilbertson (eds.), *Climate Justice and Community Renewal*, London: Routledge.

Newell, P. (2022) 'Climate Justice', *The Journal of Peasant Studies*, Vol 49 (5), 915–923, https://doi.org/10.1080/03066150.2022.2080062.

Petermann, A. (2009) 'What Is Climate Justice?', http://globaljusticeecology.org/climate-justice/.

Pickering, J. and Barry, C. (2012) 'On the Concept of Climate Debt: Its Moral and Political Value', *Critical Review of International Social and Political Philosophy*, Vol 15 (5), 667–685, https://doi.org/10.1080/13698230.2012.727311.

Preston, I., Banks, N., Hargreaves, K., Kazmierczak, A., Lucas, K., Mayne, R., Downing, C. and Street, R. (2014) *Climate Change and Social Justice: An Evidence Review*, Joseph Rowntree Foundation Report, https://www.jrf.org.uk.

Satgar, V. (2018) 'The Climate Crisis and Systemic Alternatives,' in V. Satgar (ed.), *The Climate Crisis: South African and Global Democratic Eco-Socialist Alternatives*, Johannesburg: Wits University Press, pp. 1–29.

Schlosberg, D. and Collins, L.B. (2014) 'From Environmental to Climate Justice: Climate Change and the Discourse of Environmental Justice', *WIREs Climate Change*, Vol 5 (3), 359–374, https://doi.org/10.1002/wcc.275.

United Nations (2015) *Paris Climate Agreement*, https://unfccc.int/process-and-meetings/the-paris-agreement/the-paris-agreement.

United Nations (2022) 'Secretary-General's Remarks at COP27', 17 November, https://www.un.org/sg/en/content/sg/press-encounter/2022-11-17/secretary-generals-remarks-cop27-stakeout-delivered.

Warlenius, R. (2018) 'Decolonizing the Atmosphere: The Climate Justice Movement on Climate Debt', *Journal of Environment and Development*, Vol. 27 (2), 131–155.

Wright, N.G. (2011) 'Christianity and Environmental Justice', *Cross Currents*, Vol. 61 (2), 161–190.

2 Climate Action Policies, Strategies and Programming in Southern Africa

Missed Opportunities for Climate Justice

Philani Moyo, Thabo Ndlovu, Thokozani P. Moyo, Treda Mukuhlani, Elinah Nciizah, Tendai Nciizah, Adornis D. Nciizah, Thapelo Ramalefane, Stanley O. Ehiane, France Maphosa, Grascious Maviza and Sipho F. Mamba

Introduction

In response to the persistent existential threats and human development challenges caused by the climate crisis, all countries in Southern Africa have become party to numerous international climate governance agreements, protocols and action plans. Currently, the United Nations Framework Convention on Climate Change (UNFCCC) (1992), United Nations Sustainable Development Goals (2015) and the Paris Agreement on Climate Change (2016) which is "a legally binding international treaty on climate change" (United Nations Framework Convention on Climate Change 2016:1) are the principal international agreements that Southern African countries are party to. Notwithstanding their limitations, the primary principle of these international conventions is the creation and maintenance of a global platform that coordinates mitigation, adaptation and resilience building guided by common but differentiated responsibility norms. These principles also underpin other international climate governance protocols such as the Montreal Protocol (1987), highly contentious Kyoto Protocol (1997) and ancillary climate governance treaties that include the Convention on Wetlands of International Importance (1987), Convention on Biological Diversity (1994) and Sendai Framework for Disaster Risk Reduction (2015), among others. Even though these international agreements are not perfect due to their discredited neoliberal orientation and conception of the causes, impacts and solutions to the climate crisis, their worth is in the mapping and articulation of a global climate action vision. In an attempt to actualize this global vision, Southern African countries have in their own unique ways designed, adopted, adapted and implemented numerous climate action policies, strategies and action plans with varying degrees of success and more frequent outright failure. In this chapter, we unpack and analyse selected climate laws,

DOI: 10.4324/9781003397120-3

policies, strategies and action plans of Botswana, Eswatini, Lesotho, South Africa and Zimbabwe with a view to understanding the extent to which they are pathways towards social and climate justice. By any measure, this overview is not an exhaustive analysis of all climate policies, strategies and action plans of the case study countries. Rather, a selection of the most relevant for our analysis was chosen to paint a broad picture of the state of climate injustice as the climate crisis continues to unfold.

We begin our analysis in Botswana and Eswatini. In Botswana, the country's climate agenda is primarily anchored on the National Policy on Disaster Management (1996), National Disaster Risk Reduction Strategy (2013), National Development Plan II (2017), National Climate Change Response Policy (2016) and the National Climate Change Strategy (2018), while in Eswatini, the National Climate Change Strategy and Action Plan (2013), Swaziland National Climate Change Policy (2016), National Development Plan (2019) and National Drought Plan (2020), among others, guide the national response to the devastating impacts of the climate crisis. While these climate action policies and strategies provide a clear agenda, we argue that what matters and makes a difference in the climate response is what respective governments, non-state actors, private sector, civil society and citizens do for mitigation and adaptation purposes at the local level. The local is the most important site as it is the existential space where the vulnerable, poor and marginalized are suffering procedural and distributive injustice. In that regard, our analysis demonstrates that both the Botswana and Eswatini governments have not directly problematized nor mainstreamed climate justice into their climate action policies and programming. Consequently, their grassroots mitigation and adaptation programmes do not adequately address deep-rooted issues of vulnerability, inequity and inequality without which climate justice is not attainable.

This lack of effective and sustainable practical climate actions towards climate justice is also evident in Lesotho and Zimbabwe. Despite a phalanx of policies and strategies identifying agriculture, energy, water, forestry, infrastructure and human health as sectors that are adversely affected by climate impacts, especially in poor and vulnerable communities, there are very limited and sometimes lack of budgets for mitigation and adaptation programming. As a result, the poor and vulnerable, children, women, the disabled, youth and other social groups continue to suffer climate crisis consequences with very limited opportunity of realizing social and climate justice. The same applies to South Africa where regardless of its ambitious climate action policies and strategies, it is failing to achieve greenhouse gas (GHG) emission reduction targets enunciated in its nationally determined contributions (NDCs) due to its continued reliance on coal for its energy-industrial complex. For this reason, the veneer of South Africa pursuing emission reductions as one of the actions towards climate justice is patently unrealistic. While it has made some progress on the adaptation front, specifically on enhancing grassroots climate-resilient agriculture, biodiversity and ecosystems conservation and provision of climate services, these successes are very few and fragmented, thus limiting their overall impact in transforming the lives of the poor and vulnerable. One of the ways South Africa, and the other Southern African countries under discussion here, can potentially achieve distributive and procedural justice for the poor and vulnerable is through increasing climate finance. Without adequate climate financing, many adaptation and resilience activities of the poor and marginalized will remain negligible limiting their chances of enjoying basic socioeconomic rights under the ongoing climate crisis.

Climate Justice Ambitions and Fuzzy Targets in Botswana, Eswatini and Lesotho

Similar to its peers in Southern Africa, Botswana has a plethora of climate policies, strategies and action plans. Alongside its National Policy on Disaster Management (1996), National Disaster Risk Reduction Strategy (2013) and National Development Plan II (2017), its primary climate action policy is the National Climate Change Response Policy (2016) which outlines the country's mitigation, adaptation and resilience agenda. It does so through linking national development implementation with enhancing the country's resilience capacity to respond to existing and future climate impacts (Makwatse *et al.* 2022). This can be achieved, argues the policy, through having national development actions that are informed by realization that the climate crisis is exacerbating poverty levels (Government of Botswana 2016). Therefore, to address this dual challenge of poverty and climate change, human development programming must seek to create opportunities for the poor to transition out of poverty as a path towards achieving socioeconomic justice and equity. Further, its vision of an economy that is environmentally sustainable and follows a low-carbon development pathway in pursuit of prosperity for all signals its intent of holistically factoring environmental, social and economic considerations as it seeks to build resilience. This policy agenda is complemented by the country's National Climate Change Strategy (2018) which operationalizes tenets of "the policy and provides necessary guidance on how the policy objectives will be achieved" (Government of Botswana 2018:1). In that regard, the government has enacted sectoral structures that are already implementing specific climate mitigation and adaptation actions such as government technical and financial support for solar technologies in the renewable energy sector. This demonstrates initial steps towards fulfilling the country's Paris Agreement mitigation commitment to significantly decrease GHG emissions through phased switching to renewable energy sources (Government of Botswana 2022). Relatedly, climate change information and knowledge dissemination is one of the flagship programmes of the country. This is done through a participatory process, in line with procedural justice, that involves government entities, local communities, civil society and the private sector. Alongside these government programmes, international development partners have also partnered with local communities to implement various climate action projects. These include the Collaborative Adaptation Research Initiative in Africa and Asia, Resilience in the Limpopo River Basin and the Southern Africa Regional Environmental Programme funded by the United States Agency for International Development, Adaptation at Scale in the Semi-Arid Regions funded by the Department for International Development of the United Kingdom and the International

Development Research Centre of Canada, etc. These climate actions, and many others, are evidence of initial steps that the state, non-state actors and local communities are implementing as they attempt to pursue climate-resilient development. However, as we demonstrate below, these are few and far between to create sustainable resilience.

Firstly, failure of the national climate policy and associated strategies to problematize and mainstream climate justice into policy and programming processes means one has to infer or deduce it from these climate smart programming blueprints. Their reference to public participation "to ensure that adaptation and mitigation decisions and response measures are in the best interest of the general public" (Government of Botswana 2018:10) thus becomes the basis for inferring the climate justice vision. This climate justice lacuna is striking and exposes an oversight of the policy and strategic programming direction. Secondly, the country's government-led climate actions do not adequately address deep-rooted issues of inequity and inequality without which climate justice is not attainable. It is well documented that Botswana is one of the most unequal countries in the region (Baleyte *et al.* 2021; Gordon 2019; World Bank 2020). Inequality in the country has gender, age, ethnic and rural–urban dimensions. Given the truism that low-income, poor and marginalized communities are most vulnerable to climate impacts, their climate justice should therefore be firmly linked to equality, equity and socioeconomic justice. To achieve their climate justice, it is crucial for Botswana to boldly address these underlying social injustices manifested through various forms of inequality and inequity. Without these socioeconomic changes, climate justice will remain a 'pipe-dream' because its realization partly relies on the achievement of social justice. Thirdly, while the country aspires towards a green economy, introduction of carbon budgets and carbon taxes as part of its decarbonization agenda, it does not yet have a just transition framework to seamlessly guide this transformational change. It therefore remains murky, perhaps uncertain, how the country can realistically undergo a just transition that involves "facilitating equitable access to the benefits and sharing of the costs of sustainable development such that livelihoods of all people, including the most vulnerable, are supported and enhanced..." (African Development Bank 2021:1) in the absence of a framework and consensus between local communities, civil society, public and private sector.

The policy space and alignment is slightly different in neighbouring Eswatini. The country has several laws, policies and strategies that respond to climate change in advancement of the call for climate justice and sustainable transition. For example, its National Development Strategy (2016), National Development Plan (2019) and the Strategy for Sustainable Development and Inclusive Growth 2030 are some of the macro policy and strategic documents that integrate sustainable development, environmental and climate action. Its Swaziland National Climate Change Policy (2016) provides an enabling policy framework that guides the development of a sustainable and climate-resilient society by encouraging communities, private sector and government to follow low-carbon green growth paths (Ministry of Tourism and Environmental Affairs 2016). It further raises awareness on the opportunities presented by climate change for investors and local communities to leverage on as they build climate-resilient communities and engage in poverty reduction initiatives (Ministry of Tourism and Environmental Affairs 2016). In further mainstreaming climate action in the national development agenda, climate programming features in sectoral strategies and action plans. This is the case in the National Climate Change Strategy and Action Plan, Disaster Risk Reduction National Action Plan and the National Disaster Management Policy, among others (Government of Eswatini 2005, 2006, 2007, 2008, 2009). Further, the National Drought Plan (2020), for instance, responds to the devastating impacts of recurring drought conditions (Government of Eswatini 2021). It recognizes that climate change has intensified drought conditions necessitating the need for capacity building in disaster risk reduction and climate adaptation in order to strengthen the resilience of the agricultural sector. Similarly, the country's Resilience Strategy and Action Plan provides the policy scope for upscaling the country's adaptive initiatives in order to enhance its resilience against all forms of climate hazards, shocks and damages (Government of Eswatini 2021). This plethora of policies, strategies and plans means that, on paper, Eswatini has created an enabling policy environment for climate actions towards climate justice. This is a demonstration of its endeavour to promote effective implementation of climate change adaptation and resilience in a fair and equitable manner to counter prevailing climate impacts.

However, these progressive Eswatini climate policy papers have not been complemented by effective and sustainable practical climate action. This is because the country faces several institutional and capacity constraints which deter full implementation of climate initiatives. For example, the public–private partnership between the United States Agency for International Development, National Department of Energy and the Eswatini Energy Regulatory Authority has not succeeded in setting the country on a clean, renewable energy pathway. The partnership's goal of transitioning the economy to clean energy in order to minimize GHG emissions and the national carbon footprint has not only failed because of lethargy in policy implementation but also because of inadequate climate finance to fund the transition. This inability to raise adequate climate finance from regional, continental and international sources is due to lack of requisite capacity and expertise in developing competitive and fundable proposals. Consequently, vulnerable societal groups such as the unemployed, workers, the poor, elderly and women continue to suffer from energy poverty in an environment of high GHG emissions. Secondly, while the government has made financial and technical investments in the agriculture sector through the National Agriculture Investment Plan and the continental agricultural framework – the Comprehensive African Agriculture Development Programme – these have not reduced poverty nor enhanced food and nutrition insecurity. This suggests that the country's climate smart agriculture interventions, such as the Lower Usuthu Smallholder Irrigation Project that aimed to enhance mitigation, improve household food security, in addition to the reduction of land degradation and biodiversity loss, have largely failed.

Similar conditions and evidence of climate inaction were obtained in Lesotho. This is despite the country's 1993 Constitution clearly laying the legal foundation for the protection of the environment, ecosystems and biodiversity as well as cultural resources for present and future generations (Machepha 2010). This constitutional position finds expression in, among other laws, the National Environment Act of 2008 which, despite its preoccupation with traditional environmental conservation, contains elements of climate thinking. Just like in Eswatini and Botswana, at policy and strategic levels, Lesotho has a gamut of climate action position papers. These include the National Environmental Policy (1998) that focuses less on mitigation and adaptation and more on sustainable development which was in vogue at the time of its promulgation post the seminal United Nations' 'Our Common Future' report (aka Brundtland Report). Other policies with very limited climate action insights, but more emphasis on environmental management, water, forestry, biodiversity, land degradation and desertification, include the Livestock and Range Management Policy (1994), Water Resources Management Policy (1996), National Forestry Policy (1997), National Biodiversity Strategy and Action Plan (2000), Water and Sanitation Policy (2007), National Rangeland Management Policy (2014) and the Energy Policy (2015). The glaring climate action gaps in these foregoing policies are, however, addressed through the National Climate Policy (2017) which advocates the promotion of climate-resilient social, economic and environmental development that is compatible with and mainstreamed into traditional national development plans and budgetary processes (Lesotho Meteorological Services 2017). Second, this policy provides for a just transition through pursuing low-carbon development opportunities, clean technology development, transfer and use, sustainable use of environmental resources, strengthening of climate governance and financing mechanisms in a way that benefits the most vulnerable social groups (Lesotho Meteorological Services 2017) across Lesotho.

For implementation purposes, the country's climate policy is operationalized through the National Adaptation Programmes of Action (NAPA). This NAPA is a product of a nationwide stakeholder consultation process (i.e., procedural justice), and according to Scholsberg (2004), such consultations and engagement are elements of social justice. Through its 11 adaptation activities, the NAPA deliberately seeks to empower "vulnerable communities to adopt adaptation capacities" (United Nations Climate Technology Centre and Network 2023:1). However, the direct impact of NAPA activities on adaptation and resilience of the vulnerable and marginalized remains inconsequential. This is confirmed by the government of Lesotho which acknowledges that some of its policies and programmes of action are fair, equitable and just in theory only given that the adaptive capacities of many poor communities remain precarious (Lesotho Meteorological Services 2017). This is hardly surprising because even though the national climate policy and NAPA identify agriculture, energy, water, forestry, infrastructure and human health as sectors that are adversely affected by climate impacts, there are no streamlined and funded programmes being implemented consistently to address these Anthropocene challenges. The limited and sometimes lack of budget for this programming thus adds to the ever-deepening challenges across all sectors. The corollary is that poor and vulnerable children, women, the disabled, youth and other social groups continue to suffer the consequences of the climate crisis with very limited opportunity of realizing social and climate justice.

Climate Justice Talk and Limited Action in South Africa and Zimbabwe

In many respects, South Africa and Zimbabwe are also faring badly in climate action for social and climate justice. Similar to other countries in the region, both view and approach climate mitigation, adaptation and resilience building as a human development priority in policy terms. In so doing, Zimbabwe's environmental laws that include the Constitution (2013), Forest Act (1949, as amended), Natural Resources Act (1941, as amended), Forest Act (1949, as amended), Atmospheric Pollution Prevention Act (1971, as amended), Communal Lands Act (1982, as amended), Civil Protection Act (1989, as amended) and the Environmental Management Act (2002, as amended) articulate its legal position on a range of environmental matters; some of which have a direct and indirect bearing on the climate ecosystem. These laws find environmental and climate governance expression in the National Development Strategy 1, Comprehensive Agricultural Policy Framework, National Drought Plan for Zimbabwe, Disaster Risk Management Policy and Strategy, Water Policy, Forestry Policy, National Energy Policy and the Renewable Energy Policy (Brazier 2017). The National Climate Policy (2016) and National Climate Change Response Strategy (NCCRS) (2015) are the two climate policy and strategy documents that primarily inform the national climate action agenda. Broadly, the National Climate Policy (2016) seeks to guide climate change management, climate research modelling, scale up mitigation actions, enhance national adaptive capacity, transitioning to low-carbon development, clean technology transfer and information sharing, climate education, training and awareness (Government of Zimbabwe 2016). The overarching aim of the policy is to create a pathway towards climate-resilient development and a low-carbon economy (Zhakata 2019). However, it is not alive to the discourse and actions towards climate justice and a just transition. This policy failure to capture and project how the country imagines achievement of climate justice explains why even its successive NDCs have shortcomings in relation to how the lives and livelihoods of the poor and marginalized will be improved as the country transitions to low-carbon climate-resilient development. Further, although the NDCs outline several clean energy initiatives as a way of combating GHG emissions, this remains ambitious because they do not clearly show how these will be financed – which is at the core of climate justice.

The operationalization and implementation of Zimbabwe's National Climate Policy is through the NCCRS which mainstreams mitigation and adaptation activities in economic, environmental and social development at national and sectoral levels through multi-stakeholder engagement. Although one can decipher social and climate justice thinking in the shaping of the NCCRS' adaptation pillars that focus on minimizing the socioeconomic impacts of climate change, it is, however, difficult to make a definitive conclusion about their real impact on the lives and livelihoods of the poor and marginalized at the forefront of the climate crisis. Overall, while it is evident that Zimbabwe has made significant progress in developing comprehensive policies on the climate ecosystem, many of its targets and commitments are more aspirational than practical. The country's protracted socioeconomic crisis means it simply doesn't have the financial resources to fund and implement sustainable mitigation and adaptation interventions. It therefore relies on the benevolence of international donors, specifically from the North, who have the luxury to pick and choose which project-oriented interventions to finance, for how much and for how long depending on their geostrategic interests of the time. This reliance on project climate finance adds another layer of unpredictability and vulnerability for the poor and marginalized, thus perpetuating their climate injustice. Hence, although Zimbabwe has a robust policy and strategic framework to address the challenges posed by the climate crisis, it remains uncontested that it is miles away from meaningfully addressing climate injustice.

Zimbabwe's climate action failures are surpassed by continental powerhouse South Africa which is the biggest GHG emitter in the region. Its climate governance system is legally built on the national Constitution specifically Chapter 2, Section 24 (on environmental rights, conservation and protection), and supporting laws that include, among others, the National Environmental Management Biodiversity Act of 2004, Disaster Management Amendment Act of 2002, National Energy Act of 2008, National Environmental Management: Integrated Coastal Management Act of 2008 and the Carbon Tax Act of 2019. As part of this climate governance continuum, a number of macro policies, strategies and action plans inform the national, provincial and local governments in their climate planning and programming. These include the NCCRS (2004), Long-Term Mitigation Strategy Scenarios (2008), National Climate Change Response White Paper (2011), National Development Plan 2030 (2012), National Climate Change Adaptation Strategy (2019), Green Transport Strategy (2018), Integrated Resource Plan (2019), Low-Emission Development Strategy (2020) and the Just Transition Framework (2022).

However, even though South Africa has ambitious mitigation policy and strategies, it is failing to spearhead achievement of GHG emission reduction targets in the region and to meet objectives enunciated in its NDCs under the Paris Agreement. Its absolute GHG emissions remain very high without significant progress towards reduction. One of the reasons that explains this is the country's continued reliance on coal-generated electricity through the national energy parastatal Eskom. Despite its declining electricity generation capacity due to dilapidated infrastructure and mismanagement, Eskom remains the primary coal-generated electricity producer and supplier nationally. Further, the government's appetite for coal is demonstrated by its open policy support for continued, and expanding, private sector investment in coal mining (as outlined in the Integrated Resource Plan of 2019). Some of the major South African banks are pouring millions of the local currency into new coal-mining ventures, fully supported by and in line with government policy. The grim outcome of this coal obsession is continued high GHG emissions, thus deepening of climate hazards and shocks directly affecting the poor and vulnerable. This reality means the probability of South Africa genuinely pursuing mitigation (through emission reductions) as one of the actions towards climate justice is, at the moment, patently impractical.

The lack of the South African government's urgency in pursuing mitigation actions for climate justice is also visible in its lethargic approach towards clean renewable energy generation. Despite the existence of the Integrated Resource Plan (2019) and the Just Transition Framework (2022) both of which have provisions for a gradual transition to solar, biomass, wind and small hydro energy generation, the pace of this shift is still negligible. Evidence of this abounds: while the government's efforts through the Renewable Independent Power Producer Programme (REIPPP), a private sector energy security investment initiative, is beginning to add some megawatts (MW) into the national electricity system, this is hardly a fraction of the 6,000 MW generation capacity awarded to successful bidders of the REIPPP (Government of South Africa 2023). This means the REIPPP, regardless of all the green credentials hype around it, remains an abject initiative without the capacity to produce adequate clean energy thus partly contributing to the failing mitigation agenda.

On the adaptation front, there is some progress in mainstreaming the NCCRS (2004), National Climate Change Response White Paper (2011), National Climate Change Adaptation Strategy (2019) and elements of the Just Transition Framework (2022) into provincial and local integrated development programmes. Some provincial governments (e.g., Western Cape, Eastern Cape, Gauteng, KwaZulu-Natal, etc.) and their local governments have developed and are implementing adaptation and resilience-building activities at the grassroots level. These include subsistence climate-resilient agriculture, biodiversity and ecosystems conservation, rehabilitation of wetlands, urban resilience building, urban infrastructure climate proofing, climate information sharing and climate services, etc. (Moyo 2017; Pillay and Pillay 2018). However, these are few and fragmented, thus limiting their overall impact in transforming the livelihoods of the poor and vulnerable. For example, in the Eastern Cape Province, Buffalo City's early systems warning network is poorly coordinated, hence compounding the chaotic disaster management response to recurrent climate-induced flooding in poor working-class townships. Lack of infrastructure climate proofing in these townships also means homes, roads, schools and health facilities are destroyed as and when regular flooding occurs, thus depriving the poor and vulnerable enjoyment of basic rights associated with this infrastructure.

One of the ways South Africa could potentially achieve distributive and procedural justice for the poor and marginalized, in line with its climate action strategies, is through increasing its adaptation finance purse and broadening its portfolio. In simple terms, adaptation finance "specifically targets development that reduces climate risk thereby realising climate resilience objectives" (Pillay *et al.* 2017:11). Notwithstanding that raising climate finance is difficult under the current harsh economic times in a country with lingering apartheid and colonial legacies, the fact is South Africa is simply not aggressive enough in finding these financial resources in public and private revenue sources. One of the reasons for this is the governments' approach that narrowly conceives climate finance as strictly for climate activities instead of seeing it as cross-cutting funding that simultaneously addresses other human development needs in conjunction with resilience building. This calls for the government to broaden its climate finance conception such that, for example, urban infrastructure development or rehabilitation embeds climate proofing, enhances sustainable urban living alongside strengthening resilience. For this to happen, a more dynamic and multifaceted fund-raising approach is required in order to increase the climate finance budget from its current mediocre levels. Without adequate climate financing, many adaptation and resilience activities of the poor and marginalized, whether self-initiated or not, will remain negligible and of limited impact, thus limiting their chances of enjoying basic socioeconomic rights under the ongoing climate crisis.

Conclusion

The plethora of climate policies, strategies and action plans in Southern African countries are ideologically oriented, shaped and informed by the neoliberal leaning international climate governance system. This neoliberal system, mediated through the UNFCCC and Paris Agreement, is widely criticized for unenforceable conventions and protocols as well as its lackadaisical approach in confronting geospatially unequal and inequitable impacts of the climate crisis in Africa. These weaknesses of the international climate governance system are reproduced in the five Southern African countries discussed here. In all of them, there is an abundance of blueprints on mitigation, adaptation and resilience yet little or inconsequential practical programme implementation at local level where the vulnerable, poor and marginalized continue to toil in the face of the climate crisis. What explains this lack of, or inadequate, climate-resilient development at the local level emanates from the preoccupation with developing unrealistic climate action policies and strategies. These impractical climate action plans are also divorced from the existential realities of the poor and marginalized and thus fail to address their underlying drivers of vulnerability, social and climate injustice. It is ironic that this inflated climate action ambition is not accompanied by clear prioritization of mitigation, adaptation and resilience building into mainstream national development goals. Instead, climate action is planned for, at the policy level, as an afterthought. It also doesn't help that the glossy blueprints are not accompanied by climate financing budget commitments. Without climate financing, there is no realistic possibility of these countries meaningfully transitioning to renewable energy sources, rolling out climate-resilient agriculture systems, enhancing biodiversity and ecosystems conservation, urban resilience building or climate proofing public and private infrastructure. Ultimately, it will take transformational climate leadership within these countries to place realization of social justice for the poor and vulnerable at the centre of climate action. Without such climate leadership, distributive and procedural justice will continue to elude them as long as state and non-state actors in Southern Africa continue to pay lip service to the climate crisis.

References

African Development Bank (2021) 'Just Transition Initiative to Address Climate Change in the African Context' (available at https://www.afdb.org/en/topics-and-sectors/initiatives-partnerships/climate-investment-funds-cif/just-transition-initiative).

Baleyte, J., Gethin, A., Govind, Y. and Piketty, T. (2021) 'Social Inequalities and the Politicisation of Ethnic Cleavages in Botswana, Ghana, Nigeria and Senegal, 1999–2019', in Gethin, A., Martínez-Toledano, C. and Piketty, T. (eds.), *Political Cleavages and Social Inequalities: A Study of Fifty Democracies, 1948–2020*, Cambridge, MA and London: Harvard University Press, https://doi.org/10.4159/9780674269910-018.

Brazier, A. (2017) *Climate Change in Zimbabwe: A Guide for Planners and Decision-Makers*, Harare: Konrad Adenauer Stiftung.

Gordon, R. (2019) 'Inequality, Patronage, Ethnic Politics and Decentralization in Kenya and Botswana: An Analysis of Factors that Increase the Likelihood of Ethnic Conflict', Unpublished Masters Thesis submitted to Western Michigan University (available at https:////scholarworks.wmich.edu/masters_thesis/4722).

Government of Botswana (2016) *National Climate Change Response Policy*, Gaborone: Government of Botswana.

Government of Botswana (2018) *Botswana National Climate Change Strategy 2018*, Gaborone: Government of Botswana.

Government of Botswana (2022) *Voluntary National Review Report: Botswana 2022* (available at https://hlpf.un.org/sites/default/files/vnrs/2022/VNR%202022%20Botswana%20Report.pd).

Government of Eswatini (2005) *Comprehensive Agricultural Sector Policy*, Mbabane: Ministry of Agriculture.

Government of Eswatini (2006) *National Food Security Policy for Swaziland*, Mbabane: Ministry of Agriculture.

Government of Eswatini (2007) *National Biodiversity Conservation and Management Policy*, Mbabane: Eswatini Environmental Authority.

Government of Eswatini (2008) *National Biofuels Development Strategy and Action Plan*, Mbabane: Ministry of Natural Resources and Energy.

Government of Eswatini (2009) *National Energy Policy Implementation Strategy*, Mbabane: Ministry of Natural Resources and Energy.

Government of Eswatini (2021) *Eswatini Initial Adaptation Communication to the United Nations Framework Agreement on Climate Change*, Mbabane: Ministry of Tourism and Environmental Affairs.

Government of South Africa (2023) 'Renewable Independent Power Producer Programme' (available at https://www.gov.za/about-government/government-programmes/renewable-independent-power-producer-programme).

Government of Zimbabwe (2016) *Zimbabwe National Climate Policy* (available at https://climatechange.org.zw/wp-content/uploads/2022/10/National-Climate-Policy.pdf).

Lesotho Meteorological Services (2017) *Lesotho's Nationally Determined Contribution under the United Nations Framework Convention on Climate Change*, Maseru: Ministry of Energy and Meteorology (available at https://unfccc.int/sites/default/files/NDC/2022-06/Lesotho%20First%20NDC.pdf).

Machepha, M.M. (2010) 'Parliamentary Role and Its Relationship with Its Relevant Institutions in Effectively Addressing Climate Change Issues in Lesotho', A Study Commissioned and Supported by the International Institute for Environment and Development in Partnership with European Parliamentarians for Africa (available at https://www.iied.org/g03024).

Makwatse, K., Modie, L., Mopalo, R. and Mapitsa, C.B. (2022) 'Gender Equity Considerations for Building Climate Resilience: Lessons from Rural and Peri-Urban Botswana', *Sustainability*, Vol. 14 (17) (available at https://doi.org/10.3390/su141710599).

Ministry of Tourism and Environmental Affairs (2016) *Swaziland National Climate Change Policy*, Mbabane: Ministry of Tourism and Environmental Affairs.

Moyo, P. (2017) 'Vulnerability and Assets Nexus in Climate Change Adaptation: Reflections from South Africa and Zimbabwe', in Akpan, W. and Moyo, P. (eds.), *Revisiting Environmental and Natural Resource Questions in Sub-Saharan Africa*, Newcastle: Cambridge Scholars Publishing, pp. 75–91.

Pillay, K., Aakre, S. and Torvange, A. (2017) *Mobilising Adaptation Finance in Developing Countries*, Oslo: Centre for International Climate Change Research (available at https://pub.cicero.oslo.no/cicero-xmlui/bitstream/handle/11250/2435614/Torvanger%20 2017%2002%20web.pdf?sequence=1&isAllowed=y).

Pillay, K. and Pillay, S. (2018) 'Financing Climate Adaptation and Resilience in South African Cities', in Davidson, K. (ed.), *South African Cities Network: State of City Finances Report 2018,* Johannesburg: South African Cities Network.

Scholsberg, D. (2004) 'Reconceiving Environmental Justice: Global Movements and Political Theories', *Environmental Politics,* Vol. 13 (3), pp. 517–540.

United Nations Climate Technology Centre and Network (2023) 'National Adaptation Programme of Action: Lesotho' (available at https://www.ctc-n.org/resources/national-adaptation-programme-action-lesotho).

United Nations Framework Convention on Climate Change (2016) 'Paris Agreement: What Is the Paris Agreement' (available at https://unfccc.int/process-and-meetings/the-paris-agreement).

World Bank (2020) 'World Bank Group Climate Change Action Plan 2021–2025: Supporting Green, Resilient, and Inclusive Development' (available at https://www.worldbank.org/en/publication/wdr2020).

Zhakata, W. (2019) 'Governing Climate Change: General Principles and the Paris Agreement', in Murombo, T., Dhliwayo, M. and Dhlakama, T. (eds.), *Climate Change Law in Zimbabwe: Concepts and Insights*, Harare: Konrad Adenauer Foundation, pp. 48–61.

Part II

Political Economy of Adaptation, Resilience, Injustice and Just Transition

3 Drought Adaptation Practices and Rangeland Management in Rural Umzingwane, Zimbabwe

Implications for Climate Justice

Thabo Ndlovu and Philani Moyo

Introduction

In the 21st century, climate change has become the most significant social and political challenge. As a result, communities in the Global North and South are experiencing weather-related hazards, such as floods, droughts, intense summers, extreme heat waves, slight cool weather events, and storms of increased frequency and magnitude with huge economic and social consequences. In Africa and Zimbabwe, in particular, drought has emerged as one of the significant threats to livelihoods and well-being of rural communities. Due to climate variation, droughts are now more recurrent and severe across the African continent (Nepal et al., 2021) than ever before. Consequently, drought stress compounds pre-existing environmental challenges in the dryland African savannah. These overlapping, and deepening, problems include dryland degradation, desiccation, or aridification (Darkoh, 1998) with direct effects on smallholder livestock farming. The drought risk combined with weak communal grazing systems exacerbates challenges faced by smallholder livestock farmers and accounts for differences in exposure and susceptibility to drought shocks (Adhikari and Panda, 2018). These experiences of unequal distribution of climate-induced impacts have galvanized the need to invest in climate justice following the realization that the least prepared and least likely communities to contribute to the deleterious climate effects are the most affected by climate consequences (Nicholas and Breakey, 2017).

Broadly, climate justice, a contested concept in practice, embraces principles of fairness, equity, and rightness of responses to climate change (Harris, 2019), while according to Nicholas and Breakey (2017), it denotes ethical and human rights issues that occur as a result of climate change. The principles of justice express what has to be done across the dimensions of justice. One such example is the belief that fairness exists when those contributing more towards the destruction of the environment pay more towards its redress which is a criterion of proportionality (Song and Chuenpagdee, 2015). Others hold the view that climate change contestation did not result from the intention to do harm nor consciousness of imminent consequences but blurred moral transgression and ethical violation of rights (Sovacool and Dworkin, 2014). It is a concept framed from the lens of rights, responsibilities, and procedures. Climate justice can be experienced on account of

DOI: 10.4324/9781003397120-5

differences in income and wealth, race, gender, ethnicity, age, and sexual identities within nations (Kashwan, 2021). The climate justice concept, first coined in 1989 (Schlosberg and Collins, 2014), sought to magnify accountability matters about whose voice is heard and whose interests are served in fulfilling its dimensions on mitigation (emission reductions), adaptation (tackling the impacts), and loss and damage (dealing with the residual adverse impacts after adoption of mitigation and adaptation) (Newell et al., 2020).

The concept has played a significant influence in shaping the United Nations Framework Convention on Climate Change (UNFCCC), its Kyoto Protocol, and the global treaty signed in Paris in December 2015. The Kyoto Protocol is one of the major attempts towards addressing climate change as it focuses on obligations and voices towards curbing global emissions through multilateral governance. Despite all these positives, the Kyoto Protocol has been criticized for its lack of fairness and equity. Resultantly, the Paris Climate Agreement motivated by political interest was, however, deemed an economic favourable pathway (Glanemann et al., 2020) centred on the concept of voluntarism by countries to observe emission targets. While the December 2015 Paris Climate Agreement is better than no agreement, Clémençon (2016) aver that it has abandoned equity and justice considerations as a guiding principle for multilateral cooperation. Effectively, it has let the major contributors to climate change off the hook, letting vulnerable groups such as smallholder livestock farmers to bear the brunt of the climate crisis. Debates on fairness and equity are growing louder; hence, climate justice is gaining traction in strengthening current and future anthropogenic climate change interventions largely viewed as central in reproducing or exacerbating inequities worldwide (Krause, 2018). In this context, adaptation to climate-induced droughts and climate justice needs to be intertwined within policies and legislative frameworks to avoid resilience building being an empty process that reproduces inequalities and entrench defencelessness of smallholder livestock farmers (Campell Torres et al., 2020).

As a result, this chapter examines the significance of climate justice in strengthening communal rangeland management and communal livestock farmers' adaptive capacities in the Umzingwane district against climate risk. Local grazing management systems and legislative frameworks are engaged to discern the connection between communal farmers' adaptation strategies and climate justice dimensions, namely recognitional, distributive, and procedural justice (Juhola et al., 2022). These dimensions exist to embolden vulnerable communities to demand the right to be cushioned and protected from climate injustice and tasks nations with the responsibility to absorb harm resulting from climate variation to adopt procedures that magnify their voice and efforts in addressing climate-induced shocks (Travers et al., 2019). Recognition in the justice context concerns who is given respect and whose interests and values are embraced when making decisions. Further, it entails the observance of diversity of worldviews of the vulnerable based on gender, race, religion, or ethnicity and a clear understanding of their experiences in promoting adaptation to climate-induced shocks fairly (Schlossberg and Brown, 2004). Recognition speaks to the understanding of who enjoys the benefits as well as those shouldering the burdens that might result from climate intervention processes and activities. The recognition of existing groups of farmers is crucial for improving their status and lays a basic foundation to unearth simmering conflicts and tensions that they may suffer from over land utilization and skewed natural resource management practices. This consideration shapes recognition of different groups (Walker, 2012) and is significant in the smallholder livestock farming space in lessening the dominance of other social categories through adoption of procedures which promotes equity in decision making. However, Hurlbert and Rayner (2018) suggest that recognition and simple provision of due processes may not offer sufficient attention and enough voice of the marginalized and the oppressed. Recognition without effective participation implies the involvement of local people, yet without much influence on the outcomes of the adaptation processes (Cooke and Kothari, 2001). Complementing recognition is procedural justice which entails a process that is transparent, accountable, and fair to effectively give voice and decision control to affected smallholder farming communities in a respectable manner (Ruano-Chamorro et al., 2021). This dimension is about who is involved and with what influence in terms of decision-making (Svarstad and Benjaminsen, 2020), and it draws from various power perspectives. The engagement of those exposed to climate risk presents an opportunity to address the unjust distribution of costs and benefits following the roll out of adaptation processes. In this context, this chapter contributes to adaptation discourse as it explores climate justice in communal grazing practices, informal and formal local pasture management. The grazing practices and rangeland management are critiqued in the context of existing natural resources by-laws, influence of traditional leadership, land custodianship, and regulatory authority of state institutions in strengthening the adaptation of smallholder livestock.

Communal Rangeland Management: Legislative Framework, Practices, and Climate Injustice

Land ownership patterns and rangeland management practices in communal areas of Zimbabwe can be traced back to the colonial era legacy. For example, the Land Apportionment Act of 1930 demarcated land on racial lines and dispossessed black people of their fertile land (Chivandi et al., 2010) relegating them to poor, dry infertile soils. Relatedly, to enforce the establishment of environmental protection measures, the Native Land Husbandry Act of 1951 was passed resulting in the restriction of livestock numbers owned by black people in Tribal Trust Lands (TTL), which we now know as communal areas (Thebe, 2012). The passing of pre-independence laws reflects deep-rooted traits of non-participation of the vulnerable in policies and legislation that purport to shield them from shocks. The legislative framework reflected an attitude inclined towards safeguarding the natural environment with less or no focus on the impacts of these on people or their livelihoods (Ulibarri et al., 2022). Such injustices, not people oriented, weakened the ability of smallholder farmers to adapt to climate-induced shocks (Meerow et al., 2016). In an attempt to tame land degradation, Tribal Land Authorities were established in 1970, and this gave recognition to chiefs as authorities in preserving natural resources in communal areas (Mamimine and Mandivengerei, 2001). Chiefs wielded relative power over rangeland management and grazing practices during the colonial era. The unfettered power granted to chiefs on the regulation of natural resources is testimony of decision-makers designing pseudo participation, community-based, and collaborative processes which rarely served the interest of those exposed to climate-induced drought pre-independence. The strategy by the colonial government focused less on existing inequalities and how climate change actions exacerbated or entrenched underlying structural disadvantages that deepened climate injustice.

In post-colonial Zimbabwe, agricultural land is administered through formal and informal structures viewed as oppressive and not protecting the interest of the vulnerable groups. In this sense, adaptation and climate justice must be intertwined within the planning and legislative frameworks so that resilience building does not become a reproductive model of the production of inequalities (Campello Torres et al., 2020). It is therefore crucial to promote critical interrogations into how different sociocultural identities, values, and behaviours are recognized and embedded into policy discourses (Chu and Michael, 2019). The formal system is legally anchored on the Constitution of Zimbabwe (Chapter 72:1a–b) which defines agricultural land as "land used or suitable for agriculture, that is to say for horticulture, viticulture, forestry or aquaculture or for any purpose of husbandry… but does not include Communal Land" (Government of Zimbabwe, 2013: 27–28). On the other hand, "communal land is land set aside under an Act of Parliament and held in accordance with customary law by members of a community under the leadership of a Chief" (Government of Zimbabwe, 2013: 93). This distinction of communal land from agricultural land, and its legal definition, has a number of ramifications. Firstly, the idea of communal ownership of land under customary law is a smokescreen of state ownership because the state retains control under an Act of Parliament and through its proxies which ultimately limited the rights of traditional chiefs; hence, adaptation scope of drought-exposed smallholder farmers was constrained. Secondly, and relatedly, the pseudo communal ownership means community members don't have tenure security but have fragile user rights. Thirdly, other land-related problems in communal areas include inequalities in land sizes per household, environmental degradation, and declining land productivity (Moyo, 1995; Mbiba, 2001). The question that arises is whether the exercise of political power by authorities is conscious of the disparities between diminishing grazing land and community aspirations of strengthening agriculture-related livelihoods such as livestock farming.

Political power and traditional leadership authority are central in understanding communal land management and grazing practices in post-colonial Zimbabwe. The Communal Land Act of 1982 ceded most authority of traditional leaders to Rural District Councils (Mohammed-Katerere, 2003) but allowed them to continue allocating land within their communal areas of jurisdiction. Their expanded powers, roles, and functions were later defined in the Traditional Leaders Act of 1998 which mandates them with ensuring sustainable exploitation of natural resources that includes grazing pasture, flora, and fauna (Musekiwa, 2012). Traditional leaders have power to enforce adherence to grazing schedules in consultation with their constituencies (Chigwata, 2015). Despite some of their positive roles, traditional leaders are confronted by numerous community existential challenges that include interference by political and economic elites. The disdain by elites compromises both distributional and procedural justice due to lack of respect and obedience towards local by-laws. Another challenge they and other community members face is the climate crisis and attendant frequent droughts which exacerbate poorly managed natural resources. In fact, the impacts of recurrent droughts are testing the legislative framework and local practice contribution to adaptive and resilience capacities of smallholder farmers including those in the Umzingwane district as we argue further.

In Zimbabwe, adaptation and resilience to natural hazards such as drought is guided by the Department of Civil Protection (DCP) working with the Agricultural Advisory and Rural Development Services. The DCP is empowered by the Civil Protection Act of 1989 to lead and coordinate all civil protection matters (Kudzai, 2019) and supports areas that fail to adapt. This Act, which is predominately response oriented, informs resource mobilization and administrative matters at the point of intervention (Manyena et al., 2013). However, the Act and DCP have been criticized for being outdated and reactionary instead of being proactive. The DCPs' predominant focus on civil protection and emergency management, as opposed to a holistic approach to disaster risk management that prioritizes proactiveness, is parochial. Thus, the DCP falls short in many respects; hence, some of the drought adaptation interventions meant to entrench resilience are not prioritized. While the Civil Protection Act voices the need for every citizen to act to mitigate hazards including climate induced, it has no provision to promote participation of vulnerable groups, and this deprives them space to express their aspirations and identify existing interventions in a transparent and accountable manner. The gaps in legislation compound the complexity of climate change and the framing of mechanisms for sharing burdens of climate action (Markowitz and Shariff, 2012). In light of the short falls, Zimbabwe formulated a National Climate Change Response Strategy in 2014 as part of its efforts towards the strengthening of resilience to climate variation (Ndlovu, 2022). Further, Zimbabwe ratified the UNFCCC to improve climate change adaptation despite not having a stand-alone climate change policy and legislation. In 2020, through the UN CC: Learn Programme's Southern Africa Initiative, Zimbabwe updated its National Climate Change Learning Strategy to address gaps in awareness and learning in both formal and informal settings. The development of this strategy was informed by a multistakeholder consultative process in proffering context-specific interventions and ultimately contributing to adaptation and asset creation. The emphasis on context-specific intervention borrows from the Sustainable Livelihoods Framework (SLF) significant in understanding the management of natural resources and climate justice nexus.

In unpacking local practices, legislative frameworks, and pasture management, the SLF aids the analysis of the assets, livelihoods, and transforming structures in shaping livestock productivity (Petersen and Pedersen, 2010) in climate risk settings. It explores people's existential conditions such as rights, procedures, and their skills and possessions with a view of maximizing resource allocation and exploitation for enhanced well-being and strengthened resilience. This view is corroborated by Ashley and Carney (1999) that enhancing local capacities can only be successful if the household economy is understood in the local policy context in terms of the relevance of procedures. These capacities are necessary to anchor adaptation processes. Adaptive capacity can be an unintentional process, proactive or reactive (Adger, 2006), and it mirrors the dynamics involved in managing grazing resources for the benefit of cattle (Zolli and Healy, 2012). Adapting communities adjust in times of distress for them to remain functional without experiencing major deformation in structure or form (Bene et al., 2014). Further, adaptability combines experiences, knowledge, innovativeness, and adjusting responses to meet the ever-changing internal and external processes (Folke, 2016). Within the context of rangeland management, Folke (2016) equates adaptation to implementation of local practices such as planting fodder tolerant to low precipitation as well as diversifying livelihood bases to address climate risks. Other adaptive processes include mobility of cattle from dry, depleted grazing zones to less grazed and better-conditioned areas. Following this, this chapter focuses on the methodology, results, and conclusion.

Study Area and Methodology

This study was conducted in the purposively selected Umzingwane district in Matabeleland South Province in Zimbabwe due to its proneness to climate-induced droughts, the dominance of livestock farming as a livelihood source, as well as its proximity and the low cost involved in accessing participants. The district spans agroecological regions IV and V (Mugandani et al., 2012); it receives an average annual rainfall of 400–450 mm and is characterized by infertile soils, dry spells, as well as high temperatures. The dominant landforms are bare granite hills and hills covered with vegetation, separated by areas of flat lands and occasional flat rock structures. Extensive cattle farming, at both communal and commercial scales, is practised in the district due to the dominance of favourable sweet veld. The flora and fauna in Umzingwane are typical of a dry savannah ecosystem, predominantly woodlands consisting largely of *Brachystegia spiciformis*, *Colophospermum mopane*, *Terminalia*, *Acacia* species, and *Combretum* species, while thatch grass species (*Hyparrhenia filipendula* and *Heteropogon contortus*) are common as well (Maviza and Ahmed, 2020).

The study purposively targeted five wards (1, 2, 5, 8, and 10) with a population of 3,324 communal cattle owners. These wards were selected based on the local rainfall patterns, vegetation type, livestock numbers, and accessibility. Within the wards, informant-rich respondents were targeted in order to inform the study; hence, those experienced in livestock rearing especially cattle were engaged. The characteristics of the respondents included those owning cattle, actively involved in managing cattle within the ward and at the district level. Purposive sampling targeted men and women to ensure views gathered do not segregate and reflect climate justice sentiments from particular groups. The sample size of 25 key informants was considered large enough to embrace the diversity of the characteristics within the population and to inform the study. Qualitative methods informed the data collection processes with in-depth interviews conducted to capture views on communal grazing practices, rangeland management, and legislation in climate justice settings towards strengthening adaptation capacities of smallholder farmers in the Umzingwane district. Further, five focus group discussions (with 12 members per group including women) were held at a ward level to stimulate debate and gather insights on how communal grazing practices, legislation, and management shape adaptation. The key informants were drawn from the Department of Veterinary Services, AGRITEX, Environmental Management Agency (EMA), Livestock Development Committees and Traditional Leaders, and non-governmental organizations. Data collected through in-depth interviews and focus group discussions was transcribed, and a cross-case analysis was conducted to note themes from the combined responses of all key informants. Data was processed, and major inductive themes were identified. The processed themes were analysed to distil diverse thoughts and inputs shared by key informants.

Smallholder Farmers' Drought Adaptation Strategies and Climate Justice

Communal rangeland is the major source of feed for local livestock, and its management requires astute strategies to withstand frequent and intense droughts in order to promote fair utilization of veld by smallholder farmers in rural Umzingwane. As discussed earlier, climate-induced droughts characterized by limited precipitation are a huge threat to the growth, development, and diversity of communal grazing resources. To counter the negatives of drought, smallholder farmers are subjected to regular trainings on veld management which on many occasions has not lessened the impacts experienced from climate-induced droughts as evidenced by gaps towards the strengthening of communal grazing resources. The observations on existing gaps confirm the widely held view on communal rangeland disorganization, especially the famous grazing schedules which smallholder farmers rely on to manage grazing resources. This practice has been in existence since time immemorial and is largely under the purview of the traditional leadership who possess statutory and soft power in managing common pool resources that include grazing pastures. Grazing schedules are not effective in common pool resources as those with more livestock have no incentive of rehabilitating the environment. Further, the practice often lacks transparency and accountability as those operating outside the agreed norms often benefit and significantly contribute to procedural injustice due to power dynamics within the smallholder farming space. Failure to understand the theories of power in planning and implementation of grazing schedules results in unfair distribution of risk and benefits as those adhering to the prescripts of the local arrangements and contributing less to harmful acts have limited space to voice and correct anomalies. Consequently, communal grazing resources are overutilized and vulnerable to depletion from climate-induced droughts, creating the need for supportive infrastructure to prevent animals from straying into areas reserved for future use. The disregard for schedules by farmers is a challenge as Respondent 10 noted that "some farmers are arrogant, they ignore local by-laws and graze their livestock anywhere arguing that they were free to graze wherever in an independent country", implying the exercise of one's rights without limits. In addition, Respondent 10 (summarized) went on to indicate that "the attitude exhibited by some livestock farmers was emanating from the limited knowledge on the existing grazing schedule by-laws and this created the need for awareness on sustainable grazing management and the punitive measures for defiant members".

Grazing schedules need to be cushioned from abuse by political and economic elites whose attitude mirrors the historically dominant model of development that thrived on "grow now/clean up later". The attitude promotes exploitation of resources without observing the need to provide for future generations and is against the principle of fairness as it contributes to the disproportionate distribution of benefits and harm. Smallholder farmers with limited or no livestock suffer heavily from the depletion of natural resources, and this weakens their inability to adapt. Actions by political and economic elites confirm total disregard of justice and the unequal costs deepened by their negligent behaviour. Compounding climate injustice is the lack of a common understanding by smallholder farmers and the local leadership of what is just or unjust owing to differences in moral concerns. More needs to be done in the smallholder farming settings to curb irreparable environmental damage and the perpetuation of social inequality arising from the pursuit of economic gains which takes centre stage. In essence, climate justice is considered a secondary objective in implementing climate actions (Ekins and Zenghelis, 2021). This view is also shared by Marpi and Erlangga (2021) and Pegels and Altenburg (2020); economic growth may be pursued at scales that undermine just transition which in this context speaks to managing rangelands in a manner that is sustainable and inclusive.

The disagreements on the grazing schedules reflect entrenched total disregard of local practices and by-laws by some farmers. In agreement, Respondent 7 suggested that "Smallholder farmers need to collectively enforce adherence to local grazing legislation as the majority of the farmers do as they please and compromise the quality of veld". However, one of the smallholder farmers (Respondent 5) proposed the strengthening of the by-law-guiding grazing schedules through establishing grazing management committees at the local level to enhance compliance with grazing regimes. The setting up of local grazing management structures is significant in building institutions to enforce compliance, diffuse power imbalances, and minimize investment in climate actions that exacerbate vulnerabilities of smallholder livestock farmers. The SLF is silent on power inequalities (Olivier de Sardan et al., 2017), which reflects one's ability to exert influence. Political capital is an asset that can give rise to inequity within or between individual farmers (Pasteur, 2011) if not well applied, and it does influence the connection and access to resources as well as power brokers. The disparities in accessing and usage of communal pool resources reflect a society impacted disproportionately and unfairly positioned to absorb climate-induced shocks; hence, political power should be leveraged to build collective actions that enhance well-being (Aigner et al., 2002). This makes levelling of political capital relevant in adaptation efforts as it often creates opportunities and platforms for some smallholder farmers and deprives others space to engage and make decisions that smoothen just transition on managing grazing resources for the betterment of their livestock. The adaptive capacity of communal farmers is weakened as some are not adjusting grazing behaviour in response to changing climatic circumstances. This compromises veld rejuvenation and contributes to poor cattle health and body condition which easily succumbs to climate-induced drought shocks.

In understanding the effectiveness of drought adaptation strategies, an analysis of local practices in the climate justice settings is inevitable. To this end, common practices such as establishing livestock management committees, paddocking, breeding, and destocking including constraints to veld development like bush encroachment are discussed. The grazing conditions have on numerous occasions been worsened by the use of ill-timed fires as a veld management strategy. Such adaptation actions by smallholder farmers raise questions on how risks generated can be effectively distributed in a fair and equitable manner. Unwittingly, fires have led to the growth and sprouting of invasive species; some of which thrive in drought situations. The destructive nature of fires experienced in the past informed the setting up of fire-fighting structures at village and ward levels, while at the same time, they raise questions on the understanding of climate justice in rural settings such as Umzingwane. In concurrence, Respondent 2 opined that they "learnt from the past that indiscriminate fires destroy and wastes grazing resources and disadvantages farmers without capacity to sustain supplementary feeding expenses"; hence, Village Fire Management Committees were set up to work closely with the statutory EMA. The committee disseminates awareness messages and conscientizes constituencies on the negatives of uncoordinated burning regimes and the disproportionate impact on livelihoods, especially in drought years. The committees are significant in promoting procedural justice concerns for the governance of natural resources and influencing how these procedures oversee distribution of benefits and risks (Lau et al., 2021).

Establishing local management structures magnifies the reflection of direct participation of drought-exposed communities and presents an opportunity to learn about themselves and their capabilities. The procedures adopted towards addressing climate-induced droughts are significant in attaining climate justice. Not only are procedures significant in keeping communities abreast about their context and the external influence on their circumstance, but opportunities are not to be missed, and the ability to govern themselves is outlined. Use of local structures in promoting climate justice concept is viable in promoting the leaving no-one behind mantra and in understating the disproportionate impacts of climate-induced shocks. In unison, Respondent 7 posited that they

were united in curbing indiscriminate burning of the veld as they convene on a regular basis to engage on the need to collectively pool human and material resources to combat the deterioration of grazing resources due to drought as well as to improve the conditions of their livestock.

The existence of fire management committees dovetails with SLF view on the role of transforming structures in adopting context-specific adaptation strategies towards curtailing unsustainable utilization of natural assets such as veld. The local committee is relevant in strengthening adaptation to drought by supporting locals to learn from past experiences and adjust actions in order to prioritize appropriate grazing management interventions and minimize chances of generating risks.

Further, poor paddocking in the Umzingwane district threatens rangeland management efforts by the farming community. This view was expressed by Respondent 3 as they confirmed that "communities destroyed fences and control of cattle is increasingly becoming a challenge and a threat to food security". This has made it difficult to conserve grazing pastures and protect dryland cropping from cattle damage. Lack of livestock infrastructure has resulted in the easy spread of diseases such as foot and mouth, which has on numerous periods led the Department of Veterinary Services to ban cattle movement in and out of the district. While communities are advocating for fencing to help them manage grazing resources, Sun et al. (2020) argue that it has the propensity to derail gene flow and make it difficult for confined populations to complete their whole life cycle (e.g., reproduction, migration, and foraging). Gene flow is critical in adaptation as new and tolerant traits may be suppressed. Poor cattle breeds and genetics in the area is a huge setback for farmers' adaptation to climate-induced droughts. Communal cattle farmers retain the same traditional breeds for sentimental value alone. The continued rearing of genetically poor breeds indicates failure by communal farmers to adjust and align livestock management activities with changing climate-induced drought circumstances, a practice that perpetuates vulnerability. Smallholder farmer's right to make decisions should be tested to avoid defeating the gains towards climate justice. Premising breeding on sentimental value deprives climate risk societies an opportunity to tap into livestock traits that adapt easily to their context. Further, the actions by the vulnerable strengthen the arguments on the difficulties in assigning burdens on the loss and damage compensation due to climate change. Such debates derail climate justice efforts as they become a source of both momentum and controversy (Okereke and Coventry, 2016). In addition, the shortage of genetically superior bulls in the area contributes to in-breeding, and this stifles herd diversification as suggested by Respondent 1 that "communal cattle production is at risk due to limited bulls in the area which promotes line breeding". This continues to constrain the implementation of effective breeding programmes to inject traits tolerant to the drought and diminishing rangeland conditions. Destocking is one strategy that presents an opportunity for livestock farmers to self-organize and effectively manage rangeland. Destocking is data based since it aims to maintain an equilibrium between available pasture resources and livestock numbers. This is achieved by ensuring that grazing pressure is counter-balanced by the natural regeneration of the vegetation. In communal areas, equilibrium attainment is a challenge as livestock rearing is influenced by a variety of intentions which disregard the threats of climate-induced droughts. Some farmers indicated that they independently exercised low stocking rates as a drought adaptation strategy, while others indicated that livestock numbers per household had dropped over the years, thus not necessitating active reduction of their stock. The greatest challenge is failure by institutions concerned to compute grazing resources to inform communities on the approximate livestock-carrying capacities which ultimately deprive vulnerable communities the premise to act appropriately in balancing their resources with their aspirations. Understanding the composition of natural resources helps smallholder farmers to map their assets and deploy them sustainably. Destocking has historical and colonial connotations; hence, development institutions need to recognize historical injustices emanating from the Land Husbandry Act of 1951 which limited herd sizes in smallholder livestock farming areas. Failure to observe grazing capacities permits tragedy of the commons and favours the replication of inequitable distributional patterns of benefits and harm, which York and Yazar (2022) argue that it continues to exclude climate risk communities from realizing their potential in livestock production. To this end, just and inclusive processes need to be promoted to embrace relative experiences and abilities of participating stakeholders (Simpson and Basta, 2018).

Mapping of local resources accentuates the value of unpacking the vulnerability context as enshrined in the SLF, and it lays the foundation for building context-specific capacities of smallholder farmers to mitigate drought shocks. Through in-depth discussions, Respondent 8 opined that

> the optimum livestock carrying capacities were last computed soon after independence and this has made it difficult for smallholder farmers to monitor their herd against available grazing resources. Not only is this a challenge for farmers but the incapacitated institutions meant to provide advisory services.

In resonance, Aderinto et al. (2020) suggest the need for farmers to strike a balance between livestock numbers and forage carrying capacity, meaning short-term economic interests of livestock farmers to increase herd must be aligned with sustainable ecosystem conservation interventions. As an adaptation strategy, Respondent 1 "challenged farmers to destock while acknowledging that the majority is not willing to reduce herd sizes". Destocking whether justified or not tends to carry a notion of being insensitive to traditional values hence its applicability as an adaptation strategy is close to impossible (Ndlovu & Mjimba, 2021). Respondent 1 confirmed (summarized) that for smallholder farmers to embrace destocking as an adaptation strategy, efforts must be made to improve access to markets to help derive maximum benefits and be able to restock post drought periods". Failure by the institutions concerned to take responsibility and provide basic information on grazing resources consequently excludes smallholder farmers from advocating for decisions that resonate with their aspirations (Chu and Michael, 2019) and contributes to low levels of destocking. While the blame may be placed on the supportive institutions, the cost involved may be unaffordable. The notion on cost is corroborated by Fowlie et al. (2020) that an enormous body of evidence documents indicates that low-income and/or minority communities disproportionately exposed to climate-induced shocks are without the necessary capabilities to absorb them. Strengthening the case for smallholder farming communities to embrace destocking is constrained by widening bush encroachment.

The challenges posed by bush encroachment are seriously contributing to the negative competition for pasture, soil nutrients, water, light and poor quality of graze. Bush encroachment is synonymous with an increase in density, cover, and biomass of indigenous woody or shrubby plants. Climate change effects are gradually transforming the grass and tree species composition, hence the dominance of *Lantana camara*. *Lantana camara*, an invasive alien drought-tolerant plant popularly known as "ubuhobe in the local Isindebele language", has invaded most areas of Umzingwane and reduced grazing pastures. *Dichrostachys cinerea* species commonly known as "sickle bush in English and ugagu in Isindebele" in Umzingwane is gradually spreading in the area making grazing impossible. *Dichrostachys cinerea* forms impenetrable thickets that are difficult to eliminate due to high propagation capacity and constrains the growth of vegetation underneath. As global warming is threatening smallholder farming communities adaptation, safeguarding of existing and realization of desirable futures in spite of the known and yet unknown impacts of climate change complicates the attainment of justice (Fünfgeld and Schmid, 2020). To this end, ethical decisions have to be made on the acceptable and tolerable levels of deterioration of grazing resources to curtail deepening vulnerabilities of smallholder livestock farmers. Not only does bush encroachment constrain the growth of grasses, but it also reduces the grazing space in Umzingwane. One of the respondents (Respondent 4) indicated that

bush encroachment in the area is largely driven by frequent and intense fires as well as the spread of seeds of woody species. In addition, limited contact between smallholder farmers and extension departments contributes to poor veld management which ultimately results in the deterioration of grazing resources.

The weak engagement between farmers and transforming structures, as enshrined in the SLF, stifles the exchange of information on how to tackle bush encroachment and enhance adaptive capacities to drought. Limited access to information constrains effective participation in climate action and justice. Relatedly, access and use of assets also mediate processes. Access to assets is facilitated by laws, policies, culture, and institutions, which, to a large extent, shape the adaptation processes and the exposure of individuals and communities to climate risk. This also emphasizes the interdependencies of assets and the spiralling effect on each other. Capitals are not neutral; there is interaction, interdependence, and synergy between them (Flora et al., 2015). Equally relevant in adaptation is the observance of rights, taking responsibility and adoption of context-specific procedures in addressing climate injustices. This notion is anchored on the theory of cumulative causation which suggests that a loss in one asset presents a likelihood that there will be a loss in other assets (Mrydal 1957 cited in Duffy et al., 2016). This is true when transforming structures fail and their actions exacerbate degradation of the grazing resources resulting in livestock losing condition and competitiveness on the market. Poor livestock quality fetches low prices on the market thus depriving farmers adequate income for livelihood diversification and adaptation.

With the area experiencing more droughts, bush encroachment exacerbates its effects as it reduces grazing space and renders the growth of palatable species impossible even in times when precipitation is adequate. Failure by local communities to organize themselves to address bush encroachment has reduced the carrying capacity of the land and subsequently weakened their adaptive capacity. This led to Respondent 6 suggesting that affected communities should consider clearing bush using the food for asset creation programmes rolled out by a number of non-governmental organizations supporting social protection interventions. In addition, Respondent 6 (summarized) indicated the "need for government ministries such as Environmental Management Agency and the Agricultural Advisory and Rural Development Services to collectively devise mechanisms to support smallholder livestock farmers to rehabilitate grazing areas for sustained livestock production". Bush takeover is not only peculiar to Umzingwane, but it is a condition that is threatening many other communal areas, hence the need for collective efforts.

Communal Grazing Legislative Implications on Adaptation

Since independence in 1980, different Acts of Parliament and by-laws to regulate and promote sustainable communal land utilization, livestock farming and grazing pasture management have been enacted. For example, the need for leadership in communal resource management partly led to enactment of the Traditional Leadership Act (Chapter 29:17) of 1998. Since then, traditional leaders remain relevant in grazing management and are now recognized as a key stakeholder in communal land matters in the Constitution of Zimbabwe (Chapter 13) of 2013. The Traditional Leadership Act of 1998 legislates against continuous degradation, abuse or misuse of land and natural resources in communal areas, including in Umzingwane. It further lays the legal framework for "consultations with the village assembly" on village development plans (Government of Zimbabwe, 1998) for alignment with local aspirations on resource deployment. However, community perceptions that traditional leaders are actively involved in political power matters contribute to weak engagement and community organization which often compromises the discharge of their duties, especially the enforcement of grazing schedules. Another law, the Communal Land Act (Chapter 20:04) of 1982, was initially enacted to aid the classification of land, formulate "by-laws relating to the functions conferred" (Government of Zimbabwe, 1982) and to regulate occupation of communal land and utilization of its resources. However, one of the limitations of this Act is that it grants the president of the country and the minister concerned vast powers over communal land. These powers allow the president and the minister to make decisions without the representation from the smallholder farmers impacted by climate change. Such acts disregard the equitable impacts of climate action.

Of significance is the recognition of distributional justice by communal grazing frameworks through enforcement of adherence to legislated processes as failure compromises the attribute of fairness in arriving at decisions (Lau et al., 2021).

The legislative framework acknowledges and excludes crucial voices from the urgent, essential debates on how drought-exposed smallholder farmers can effectively adapt to changing circumstances quickly, efficiently, and with equity. While there is public involvement in environmental regulation in Umzingwane, community engagement often comes late in the process, and this makes it hard to substantively change the outcome. In light of such scenarios, Ulibarri et al. (2022) aver that communities faced with climate risk challenges need to be democratically engaged to create opportunities for public education, which would strengthen their capacity to engage in decision-making. Without constructive engagement, distributive injustice will continue to prevail and increase exposure and sensitivity of smallholder farmers to climate risk. Worsening the plight of smallholder farmers and deepening climate injustice is the misuse of political power in allocating and reallocating existing land which ultimately excludes deserving vulnerable cases and widens susceptibility to shocks. The abuse of political power has seen a rise in wanton evictions of communal households by the government by disregarding other laws that permitted the allocation of land to these communal households in the first instance. In addition, the fact that communal farmers are de facto land owners (under this piece of legislation) but with no absolute control of the land makes grazing management chaotic (Lesoli, 2012) sometimes. This legal "puzzle" has resulted in some Umzingwane farmers either deliberately ignoring or silently resisting sections of the law and other by-laws meant to regulate grazing management within the district. This further raises questions about the practicality, even desirability of legislation to control rangeland management in communal areas compared to adherence to locally preferred practices. The legislative framework weakens the resolve of communities to collectively institute adaptive measures without experiencing disturbances from other members of the society.

Other challenges on communal grazing management emanate from land ownership law (Communal Land Act of 1982 and Constitution of Zimbabwe, 2013) which stipulates that communal land is owned by the state, and user rights of communal farmers are equal. Progressive legislation that regulates management of communal pool resources and the general environment such as the Environmental Management Act (EMA) of 2002 must be advocated for and fully operationalized in smallholder farming settings. The EMA, if fully implemented, is a "game changer" as it "provides for the sustainable management of natural resources and protection of the environment" (Government of Zimbabwe, 2002) in partnership with traditional leaders, the local community, and the national government. In concurrence, Respondent 9 (summarized) suggested that livestock farmers need to collectively monitor livelihood practices so as not to compromise and degrade the environment which act as a buffer against challenging events such as drought. To adapt to climate-induced droughts, Umzingwane communal farmers should embrace the idea of balancing available grazing resources, livestock numbers, existing sustainable practices, and legislation in order to build long-term resilience.

Conclusion

Communal rangeland in Umzingwane is under stress due to the ongoing climate crisis. In response, smallholder livestock farmers rely on grazing schedules to manage their ever-dwindling pastures. These adaptation practices, although not novel, ignore the dimensions of climate justice, namely recognition, distributive, and procedural. The power dynamics in smallholder farming areas compounds the management challenges experienced with traditional leaders occasionally using statutory and soft power in managing common pool resources. Power dynamics are central in the utilization of grazing resources and ultimately widen distributive injustice and compromise the enforcement of natural resource by-laws by traditional leaders and consequently erode the adaptive capacities of smallholder farmers. The existing legislative framework is also an obstacle to effective adaptation by communal livestock farmers as it derails adaptation efforts and the attainment of climate justice. For example, the Traditional Leadership Act of 1998 grants the president and responsible minister unrestrained powers over communal land, and this impairs honest and fair engagement with climate risk-exposed farmers and reduces their responsive capacities against shocks. However, other pieces of legislation such as the EMA of 2002 are constructive and support farmers' drought adaptation through the establishment of environmental management committees at ward and village levels for sustainable utilization and protection of natural resources that include communal grazing land. Going forward, efforts should be made to educate the farming community on climate justice to avoid adoption of local solutions that perpetuate vulnerability to climate-induced shocks.

References

Aderinto, R.F., Ortega-S, J.A., Anoruo, A.O., Machen, R. and Turner, B.L. (2020), "Can the tragedy of the commons be avoided in common-pool forage resource systems? An application to small-holder herding in the semi-arid grazing lands of Nigeria", *Sustainability*, *12*(15), https://doi.org/10.3390/su12155947.

Adger, W.N. (2006), "Vulnerability", *Global Environmental Change*, *16*, pp. 268–281.

Adhikari, K. and Panda, R.K. (2018), "Users' information privacy concerns and privacy protection behaviors in social networks", *Journal of Global Marketing*, *31*(2), pp. 96–110.

Aigner, S., Raymond, V. and Smidt, L. (2002), ""Whole community organizing" for the 21st century", *Community Development*, *33*(1), pp. 86–106.

Ashley, C. and Carney, D. (1999), "Sustainable livelihoods: lessons from early experience", Department for International Development, London.

Bene, C., Newsham, A., Daves, M., Ulrichs, M. and Godfrey-Wood, R. (2014), "Resilience, poverty and development", *Journal of International Development, 26*, pp. 598–623.

Campello Torres, P.H., Leonel, A.L., de Araujo, G.P. and Jacobi, P.R. (2020), "Is the Brazilian national climate change adaptation plan addressing inequality? Climate and environmental justice in a global South perspective", *Environmental Justice*, pp. 42–46, http://doi.org/10.1089/env.2019.0043.

Chivandi, E., Fushai, F. and Masaka, J. (2010), "Land ownership and range resources management in Zimbabwe: a historical review", *Midlands State University Journal of Science, Agriculture and Technology*, *2*(1), pp. 13–25.

Chu, E. and Michael, K. (2019), "Recognition in urban climate justice: marginality and exclusion of migrants in Indian cities", *Environment and Urbanization, 31*(1), pp. 139–156.

Clémençon, R. (2016), "The two sides of the Paris climate agreement: dismal failure or historic breakthrough?", *The Journal of Environment & Development, 25*(1), pp. 3–24.

Cooke, B. and Kothari, U. (2001), *Participation: The New Tyranny?*, London: Zed Books.

Darkoh, M.B.K. (1998), 'The nature, causes and consequences of desertification in the drylands of Africa', *Land Degradation and Development, 9*(1), https://doi.org/10.1002/(SICI)1099-145X(199801/02)9:1<1::AID-LDR263>3.0.CO;2-8.

Duffy, L.N., Kline, C., Swanson, J.R., Best, M. and McKinnon, H. (2016), "Community development through agroecotourism in Cuba: an application of the community capitals framework", *Journal of Ecotourism*, https://doi.org/10.1080/14724049.2016.1218498.

Ekins, P. and Zenghelis, D. (2021), "The costs and benefits of environmental sustainability", *Sustainability Science, 16*(3), pp. 949–965, https://doi.org/10.1007/s11625-021-00910-5.

Flora, C.B., Flora, J.L. and Gasteyer, S. (2015), *Rural Communities: Legacy + Change*, Boulder, CO: Westview Press.

Folke, C. (2016), ""Resilience" of the Oxford research encyclopedia of environmental science", *Ecology and Society, 21*(4), p. 44, https://doi.org/10.5751/ES-09088-210444.

Fowlie, M., Walker, R. and Wooley, D. (2020), "Climate policy, environmental justice, and local air pollution", *Economic Studies* at Brookings Research Report, pp. 1-27, https://www.brookings.edu

Fünfgeld, H. and Schmid, B. (2020), "Justice in climate change adaptation planning: conceptual perspectives on emergent praxis", *Geographica Helvetica, 75*(4), pp. 437–449.

Glanemann, N., Willner, S.N. and Levermann, A. (2020), "Paris Climate Agreement passes the cost-benefit test", *Nature Communications, 11*(1), p. 110.

Government of Zimbabwe (1982), Communal Land Act (Chapter 20:04), Harare: Government Printers.

Government of Zimbabwe (1998), Traditional Leadership Act, Harare: Government Printers.

Government of Zimbabwe (2002), Environmental Management Act, Harare: Government Printers.

Government of Zimbabwe (2013), *Constitution of Zimbabwe*, Harare: Government Printers.

Harris, P. (ed.) (2019), *A Research Agenda for Climate Justice*, Cheltenham: Edward Elgar.

Hurlbert, M. and Rayner, J. (2018), "Reconciling power, relations, and processes: the role of recognition in the achievement of energy justice for aboriginal people", *Applied Energy, 228*, pp. 1320–1327.

Juhola, S., Heikkinen, M., Pietilä, T., Groundstroem, F. and Käyhkö, J. (2022), "Connecting climate justice and adaptation planning: an adaptation justice index", *Environmental Science & Policy, 136*, pp. 609–619.

Kashwan, P. (2021), "Climate justice in the global North: an introduction", *Case Studies in the Environment, 5*(1), https://doi.org/10.1525/cse.2021.1125003.

Krause, D. (2018), 'Transformative approaches to address climate change and achieve climate justice', in Jafry T (ed.) *Routledge Handbook of Climate Justice*, London: Routledge, https://doi.org/10.4324/9781315537689-37.

Kudzai, C. (2019), 'Cyclone Idai in Zimbabwe: an analysis of policy implications for post-disaster institutional development to strengthen disaster risk management', Briefing Paper, Oxfam, https://doi.org.10.21201/2019.5273.

Lau, J.D., Gurney, G.G. and Cinner, J. (2021), "Environmental justice in coastal systems: perspectives from communities confronting change", *Global Environmental Change, 66*, https://doi.org/10.1016/j.gloenvcha.2020.102208.

Lesoli, M.S. (2012), "Characterisation of communal rangeland degradation and evaluation of vegetation restoration techniques in the Eastern Cape, South Africa", PhD Thesis, Unpublished. Alice: University of Fort Hare, South Africa.

Mamimine, P.W. and Mandivengerei, S. (2001), "Traditional and modern institutions of governance in community based natural resource management", Commons Southern Africa Occasional Paper Series, No.5, https://opendocs.ids.ac.uk/opendocs/handle/20.500.12413/4633, Harare: Centre for Applied Social Sciences.

Manyena, S.B., Mavhura, E., Muzenda, C. and Mabaso, E. (2013), "Disaster risk reduction legislations: is there a move from events to processes?", *Global environmental change*, *23*(6), https://doi.org/10.1016/j.gloenvcha.2013.07.027.

Markowitz, E. and Shariff, A. (2012), "Climate change and moral judgement", *Nature Climate Change*, 2, pp. 242–247.

Marpi, Y. and Erlangga, E. (2021), "The criticism of social justice in economic gap", *Insignia: Journal of International Relations*, pp. 23–31, https://doi.org/10.20884/1.ins.2021.0.0.3759.

Maviza, A. and Ahmed, F. (2020), "Analysis of past and future multi-temporal land use and land cover changes in the semi-arid Upper-Mzingwane sub-catchment in the Matabeleland south province of Zimbabwe", *International Journal of Remote Sensing*, *41*(14), pp. 5206–5227.

Mbiba, B. (2001), 'Communal land rights in Zimbabwe as state sanction and social control: a narrative', *Africa: Journal of the International African Institute*, *71*(3), pp. 426–448.

Meerow, S., Newell, J.P. and Stults, M. (2016), "Defining urban resilience: a review", *Landscape and Urban Planning*, *147*, pp. 38–49.

Mohammed-Katerere, J. (2003), "Participatory natural resources management in the communal lands of Zimbabwe: what role for customary law?", *African Studies Quarterly*, *5*(3), http://web.africa.ufl.edu/asq/v5/v5i3a7.htm.

Moyo, S. (1995), *The Land Question in Zimbabwe*, Harare: Sapes Books.

Mugandani, R., Wuta, M., Makarau, A. and Chipindu, B. (2012), "Re-classification of agro-ecological regions of Zimbabwe in conformity with climate variability and change", *African Crop Science Journal*, *20*(2), pp. 361–369.

Musekiwa, N. (2012). "The role of local authorities in democratic transition", in Masunungure, E. and Sumba, J. (eds) *Democratic Transition*, Harare: Weaver Press, pp. 230–251.

Nepal, S., Neupane, N., Belbase, D., Pandey, V.P. and Mukherji, A. (2021), "Achieving water security in Nepal through unravelling the water-energy-agriculture nexus", *International Journal of Water Resources Development*, *37*(1), pp. 67–93.

Ndlovu, T. (2022), "Natural hazards governance in Zimbabwe", *Natural Hazard Science*, https://doi.org/10.1093/acrefore/9780199389407.013.445

Ndlovu, T. and Mjimba, V. (2021), "Drought risk-reduction and gender dynamics in communal cattle farming in southern Zimbabwe", *International Journal of Disaster Risk Reduction*, 58, https://doi.org/10.1016/j.ijdrr.2021.102203.

Newell, P., Srivastava, S., Naess, L.O., Torres Contreras, G.A. and Price, R. (2020), *Towards Transformative Climate Justice: Key Challenges and Future Directions for Research*, Brighton: Institute of Development Studies.

Nicholas, P.K. and Breakey, S. (2017), "Climate change, climate justice, and environmental health: implications for the nursing profession", *Journal of Nursing Scholarship*, *49*(6), pp. 606–616.

Okereke, C. and Coventry, P. (2016), "Climate justice and the international regime: before, during, and after Paris", *Wiley Interdisciplinary Reviews: Climate Change*, *7*(6), pp. 834–851.

Olivier de Sardan, J.P., Diarra A. and Moha, M. (2017), "Travelling models and the challenge of pragmatic contexts and practical norms: the case of maternal health", *Health Research Policy and Systems*, *15*(1), pp. 71–87.

Pasteur, K. (2011), *From Vulnerability to Resilience: A Framework for Analysis and Action to Build Community Resilience*. Rugby: Practical Action Publishing.

Pegels, A. and Altenburg, T. (2020), "Latecomer development in a "greening" world: introduction to the special issue", *World Development*, *135*, p. 105084, https://doi.org/10.1016/j.worlddev.2020.105084.

Petersen, E.K. and Pedersen, M.L. (2010), *The Sustainable Livelihoods Approach*, Aarhus: Institute of Biology, University of Aarhus.

Ruano-Chamorro, C., Gurney, G.G. and Cinner, J.E. (2021), "Advancing procedural justice in conservation", *Conservation Letters*, https://doi.org/10.1111/conl.12861.

Schlosberg, D. and Collins, L. (2014), 'From environmental to climate justice: climate change and the discourse of environmental justice', *Wiley Interdisciplinary Reviews: Climate Change*, *5*(3), pp. 359–374.

Schlossberg, M. and Brown, N. (2004), "Comparing transit-oriented development sites by walkability indicators", *Transportation Research Record*, *1887*(1), pp. 34–42.

Simpson, N.P. and Basta, C. (2018), "Sufficiently capable for effective participation in environmental impact assessment?", *Environmental Impact Assessment Review*, *70*, pp. 57–70, https://doi.org/10.1016/j.eiar.2018.03.004.

Song, A.M. and Chuenpagdee, R. (2015), "Eliciting values and principles of fishery stakeholders in South Korea: a methodological exploration", *Society & Natural Resources*, *28*(10), pp. 1075–1091.

Sovacool, B. and Dworkin, M. (2014), *Global Energy Justice: Problems, Principles and Practices*, Cambridge: CUP.

Sun, J., Liu, M., Fu, B., Kemp, D., Zhao, W., Liu, G., Han, G., Wilkes, A., Lu, X., Chen, Y. and Cheng, G. (2020), "Reconsidering the efficiency of grazing exclusion using fences on the Tibetan Plateau", *Science Bulletin*, *65*(16), pp. 1405–1414.

Svarstad, H. and Benjaminsen, T.A. (2020), "Reading radical environmental justice through a political ecology lens", *Geoforum*, *108*, pp. 1–11.

Thebe, V. (2012), "New realities' and tenure reforms: land-use in worker peasant communities of south-western Zimbabwe (1940s–2006)", *Journal of Contemporary African Studies*, *30*(1), pp. 99–117.

Travers, J., Schenk, E., Rosa, W. and Nicholas, P. (2019), "Climate change, climate justice, and a call for action", https://digitalcommons.psjhealth.org/publications/1279.

Ulibarri, N., Figueroa, O.P. and Grant, A. (2022), "Barriers and opportunities to incorporating environmental justice in the National Environmental Policy Act", *Environmental Impact Assessment Review*, *97*, https://doi.org/10.1016/j.eiar.2022.106880.

Walker, G. (2012), *Environmental Justice: Concepts, Evidence and Politics*, London: Routledge.

York, A. and Yazar, M. (2022), "Leveraging shadow networks for procedural justice", *Current Opinion in Environmental Sustainability*, *57*, https://doi.org/10.1016/j.cosust.2022.101190.

Zolli, A. and Healy, A.M. (2012), *Resilience: Why Things Bounce Back*, New York: Free Press.

4 "Making Little Go Far in the Context of Climate Change"

Managing Water Demand in Bulawayo, Zimbabwe

Treda Mukuhlani

Introduction

Due to climate change, water insecurity remains a challenge across the globe. Climate change has increased the magnitude and frequency of precipitation-related disasters such as droughts and floods (FAO, 2015). As a result, droughts have put a lot of pressure on existing water sources. Recent data show that two billion people live in countries experiencing high water stress, and 40% of the world's population is affected by water scarcity (UN, 2018). Moreover, 12 of the 17 most water-stressed countries are found in the Middle East and North Africa because the region is hot and dry, so the water supply is very low. As a result, growing water demands have pushed countries further into extreme stress (Rutger et al., 2019). Although most attention has been placed on North African countries, other African regions have recently started to experience water insecurity. This is accentuated by prolonged droughts, which are a consequence of climate change. Within the African continent, Southern Africa has the most significant number of countries with water shortages because of its climatology (Nomedoe et al., 2006). For example, South Africa is a water-scarce country, where the water demand is more than natural water availability in several river basins. The effects of variable rainfall patterns and different climatic regimes are also compounded by high evaporation rates across the country. The country experiences alternating periods of droughts and floods, which affects the amount of water throughout the whole country. For instance, in 2018, Cape Town, one of South Africa's most significant cities, began to dry up, and the countdown to the day the water ran out was dubbed "Day Zero" (Alexander, 2019).

Zimbabwe is among other countries in Southern Africa that has experienced water insecurity due to climate change. During the last 20 years, the country experienced periods of severe droughts that have increased the pressure on the available freshwater resources. Recent severe drought episodes were observed in 2001/2002, 2004/2005, 2006/2007, 2011/2012, 2012/2013, 2015/2016, and 2018/2019 (USAID, 2020). Therefore, many urban areas in Zimbabwe such as Harare, Bulawayo, and Chitungwiza are increasingly confronted with water insecurity.

For the City of Bulawayo, the second-largest and oldest city in Zimbabwe, water insecurity is more pronounced given that the city is located in a perennially arid

DOI: 10.4324/9781003397120-6

area in which water is often a scarce resource (Moyo et al., 2005). The city receives an average rainfall of 460 mm/year (Mkandla et al., 2005). It is physically located in the Gwayi and Mzingwane catchment area, with 665,940 inhabitants (Zimstat, 2022) constituting 5.4% of the country's population. The city mainly relies on surface water from six dams for its supply. The primary sources of water for the City of Bulawayo is surface water from the five dams which are Mzingwane, Inyakuni, Upper and Lower Ncema, and Insiza dams. Khami Dam, which is located to the North West of the city, has not been used for many years due to poor quality water as a result of wastewater discharges into the dam (Mabiza, 2013). Surface water is augmented at a minimal level with ground water. The water sources' location is outside the boundary of the city. The city is mandated to be a service provider to areas previously outside its jurisdiction, such as Mguza.

Climate change has worsened the water situation in the City of Bulawayo. This has been aggravated by extreme droughts hitting the city and the wider Matabeleland Province. The city has been experiencing below normal rainfall and recurring droughts for the past three decades due to climate change, negatively affecting its water sources. The low inflows the city's dams received due to insignificant rainfall mean that water levels in the dams remain critically low. More intense droughts affecting an increasing number of people have been linked to higher temperatures and decreased precipitation. Drought affects both dams (due to high temperatures and evaporation) and boreholes (because of a low water table). Over the years, the total inflow of major dams of Bulawayo has not matched domestic demand from one rainy season to another. This comes on the backdrop of declining water resource availability (drastic reduction in water supply dam levels) which has in the past decades led to increased water stress in Bulawayo urban water supplies (Sibanda et al., 2009). This has resulted in persistent, chronic water shortages marked by increased and more frequent water rationing by the Bulawayo City Council (BCC), with many suburbs such as Entumbane, Njube, and Emganwini getting piped-water supply once a week in certain instances (Maviza et al., 2022). Two major concepts, climate change justice and just transition, were used to analyse the City of Bulawayo's water crisis and the strategies used to address the issue. Climate change justice explores the effects of water scarcity to water users, in particular the urban poor, women, the old, and the disabled. Just transition is used to assess whether water demand management strategies being employed by the BCC are just and fair. Against this situation, climate change justice advocates for the inclusion and protection of the rights of those most vulnerable to the effects of climate change.

Coupled with unfavourable hydro-climatic conditions and geographical location is the issue of a growing population which is not being matched with a corresponding expansion in the city's water supply infrastructure. Bulawayo has witnessed rapid urbanisation from the colonial period up to the present. It should be noted that the 1982 census put the city's population at 413,814 (Musemwa et al., 2014), and in 2022 the city had 665,940 people (Zimstat, 2022) Paradoxically, the council uses water infrastructure installed during the colonial era and meant to service fewer people than the city's current population levels. The population in Bulawayo has increased twofold since 1980, but they are still using the same dams, reservoirs, and water pipelines installed during the colonial era (before 1980). The infrastructure no longer matches the current demand and is a contributing factor to water scarcity. The failure by policymakers (national and local governments) to plan, fund, and manage mechanisms to deliver water in proportion with urban growth rate represents one of the most serious threats to the city's future sustainable development.

Climate change factors are also intertwined with political issues which have made the Bulawayo water crisis persist for many decades. Under climate justice, political issues influence how communities adapt or mitigate climate change. There has been severe tension between the central government and BCC. The former was unwilling to help the City of Bulawayo find lasting solutions to its quest for sustainable water sources because of the government's perceptions of the city as a bastion of opposition politics. One of the enduring legacies of the local government in Zimbabwe has been the deep politicisation of municipalities (Mapuva & Takabika, 2020). Unlike Harare and other small towns whose water system was taken by Zimbabwe National Water Authority (ZINWA), the BCC did not allow ZINWA to take over. Political tensions over water management in Bulawayo have been well documented by scholars such as Musemwa (2014) and Chilunjika and Zhou (2013). There is political meddling from the central government (ruling party) in the affairs of the BCC, which an opposition party runs. So in some instances, political differences interfere in the progress of project implementation and water service provision. The 1998 Water Act took over water resources management from the BCC and gave it to ZINWA. According to Sibanda (2018), Zimbabwe African National Union Patriotic Front (ZANU PF) hegemony was deployed to deny the people of Bulawayo access to water as a punitive measure against the city that was rebellious to its political dominance. In addition, the city was being punished for supporting opposition political parties such as the Zimbabwe African People's Union (ZAPU) and Movement for Democratic Change (MDC). In Bulawayo, water has been used as a political campaign tool to lure votes. During elections, residents are promised massive projects which will solve the water problem, such as the Matabeleland Zambezi Water Project, and after elections, the central government goes quiet. Therefore, water is being used as a weapon against political opponents, and water insecurity remains a severe problem in Bulawayo.

In addition to climate change, the water situation in Bulawayo has been worsened by the national economic crisis. Effective responses to climate change have been limited by insufficient funding. In the late 1990s, the political and economic crisis meant the Zimbabwean dollar collapsed, and by 2008, five million people needed emergency food assistance. The collapse of the Zimbabwean dollar also meant that Water, Sanitation and Hygiene (WASH) service providers, for city councils such as Bulawayo and ZINWA, lacked foreign exchange, severely limiting the supply of spare parts and chemicals (UNICEF, 2019). Also, general distrust between residents and local authorities led to low willingness to pay (compounded by the economic crisis). As revenue from tariffs (from industry and households) declined, urban councils could not generate sufficient revenue to provide services. There was minimal/no central government budget allocation: local and national government authorities could not borrow due to arrears on existing loans and because lending agencies ceased operations. Most of these chemicals to treat water are imported and very expensive, creating a considerable challenge for a country facing severe foreign currency shortages (Makwara, 2012). There is also not enough funding from the central government to build more dams and funds to repair and replace water infrastructure. Climate justice advocates for climate finance to be used to cushion communities from the negative impacts of climate change.

Within this background and context, this chapter examines the BCC's strategies in managing water demand during times of water scarcity. Can these measures be a suitable and just approach to Bulawayo's water insecurity during the times of climate change, or the city needs practical and bold interventions such as building more water storage and using alternative sources? Is there enough water for drinking, livelihoods, and health services during times of climate change? Thus, this chapter interrogates how the BCC is managing water demand during times of water scarcity and climate change and the extent to which the BCC's water demand management strategies are achieving water security.

Climate Justice and Water Security

Climate justice is used and defined in different ways but primarily is mobilised to contest the unequal impacts of climate change both geographically and socially. It is fundamentally inequitable and unjust, and the effects are prone to geographical location of people. Therefore, climate justice advocates for prevention and mitigation inequalities caused by climate change. Climate change is adding to the layers of injustices people already face in their everyday lives. It is dependent on levels of economic development, infrastructural preparedness, social equality, politics, investments, and gender issues (Markkanen & Anger-Kraavi, 2019).

According to Bhatasara and Nyamwanza (2022), climate justice in Zimbabwe is not explicitly a question being tackled in the country. The government of Zimbabwe itself is not entirely projustice regarding adaptation to climate change in local communities. Many rural farmers have been unjustly exposed to cbrop failures and death of livestock, a consequence of lack of funding and the high investment costs for building dams and setting up irrigation infrastructure (Bhatasara and Nyamwanza, 2022). Climate change threatens the enjoyment of a range of human rights to food, health, housing, culture, development, and among them the human right to clean drinking water.

There is one human right affected more by climate change and is central to other human rights which is human right to water. The nature of injustice taking place is that the same people who lack access to water and sanitation are usually the ones most vulnerable to the effects of climate change and the least responsible for causing it in the first place (UN, 2022). Most people suffer from water insecurity. Water security which refers to "adequate quantities of acceptable quality water for sustaining livelihoods, human well-being and socio-economic development for ensuring protection against water-borne pollution and water related disasters and for preserving ecosystems in a climate of peace and political stability" (UN, 2013: 1). In developing countries such as Zimbabwe, there has been a lot of empty political rhetoric with little initiatives on climate action. Most cities such as Bulawayo still experience water insecurity. What is needed is climate action from the government and other stakeholders (the private sector, non-governmental organisations [NGOs], and the residents). Climate change activists have been calling for political leaders to ensure that their climate action plans tackle inequality, poverty, and injustice and promote the implementation of human rights (Porter et al., 2020). Communities that are located in fragile ecosystems are affected first and worst. Most women, children, and those living in extreme poverty are usually left behind and bear the brunt of increasing water scarcity and poverty. As a result, these marginalised populations face a vicious and unjust cycle in which a lack of access to water and sanitation is aggravated by extreme weather events, leading to more expensive and unaffordable services.

But where the problem starts may also be where the solution begins. An approach that guarantees the human right to water by tackling inequalities and putting people's needs front and centre – especially the needs of those whose voices continue to be marginalised and disregarded. This is both a necessary response and a step towards ending the climate change crisis. Solutions must promote equity, assure access to basic resources, and ensure that young people can live, learn, play, and work in a healthy and clean environment. It offers benefits for mitigation (stopping climate change) and adaptation (adjusting to the new normal).

Just Transition and Water Security

In response to challenges caused by climate change, like water scarcity, just transition must be factored into dealing with the issues. Just transition issues are on top of the agenda of the Paris Agreement. According to Jenkins et al., the birth of the concept of "Just Transition" emanated during the 2018 24th Conference of the Parties (COP24) to the United Nations Framework Convention on Climate Change meeting in Katowice, Poland, and was adopted under the "Solidarity and Just Transitions Silesia Declaration". Under this meeting, the Heads of State and Government committed themselves to seriously taking the impact of climate change and climate change policy on workers and surrounding communities. Just transition issues do not focus on water security issues only but on changes within the coal industry, employment and livelihoods, and climate financing. In other terms, "a just transition involves maximising the social and economic opportunities of climate action while minimising and carefully managing any challenges – including through effective social dialogue among all groups impacted, and respect for fundamental labour principles and rights" (ILO, 2019: 1).

The aspiration for a just transition depends on domestic climate policy outlined in each state's nationally defined priorities. Beukman and Reeler (2021) came up with principles which should be considered for a just transition in ensuring water security within the South African context. This chapter used the same South African principles to assess the extent to which they are being applied to ensure a just transition within the Bulawayo context. These are equity and access, inclusivity, economic efficiency, environmental and ecological sustainability, and climate investments and funding.

To ensure that there is a just transition, climate change mitigation and adaptation strategies must make sure that there's inclusivity. This implies that the needs and interests of marginalised groups like the youth and women don't seem to be unnoticed. Women must not be sidelined in issues to do with climate adaptation, and their needs must be prioritised. The effectiveness of any action to scale back the impacts of climate change requires an understanding of those gender-differentiated impacts, vulnerabilities, and capacities, to handle the particular needs of women and men. Livelihoods for those dependent upon water must not be jeopardised during transitioning.

Also, communities plagued by poverty prevalence are susceptible to water challenges as they have few resources to cushion themselves during times of water scarcity. The impacts of climate change are not homogeneous since the poorest countries and communities will be more liable to the impacts. Concerted efforts by various stakeholders to be inclusive are required to ensure opportunities are created in management and development, governance, access to opportunities, and decision-making on planning and investments in water and natural resources, while ensuring reduced vulnerabilities. Newell and Mulvaney (2013) pointed out that "in policy terms, the decision for just transition is usually directed to states".

Another key principle of just transition in the water sector is the issue of equity and access. It is the basic right for all people to have access to water of adequate quantity and quality. There must also be economic efficiency in water use because of the increasing scarcity of water, financial resources, and increasing demands; water must be used with maximum possible efficiency. Economic efficiency principles must be more prominent in water allocation between sectors and within sectors. Allocation efficiency must inform development planning and choices and water allocating between energy and agriculture. The movement of the just transition requires adopting activities aligned with environmental and ecological sustainability. Additionally to economic efficiency calculations, water allocations must include strategic decisions regarding the worth of sectors to the longer term low-carbon economy, including the need to sunset certain sectors with high environmental footprints and low potential future amount.

One of the foremost important to contemplate is the issue of funding as investments in climate finance are needed to make infrastructure, institutions, and data dissemination. Building engineering solutions are vitally important for water security but will not be enough to resolve the world's water problems. A wide range of social, economic, and political challenges must be addressed, requiring an equally broad selection of hardware and software tools through which this may be done. A water-secure world necessitates investment in better and more accessible information, robust and adaptable institutions, and natural and man-made infrastructure to store, transport, and treat water.

Methodology

This study utilises a qualitative approach which allows for an inductive process (interviews and document analysis) which informs interpretation. This approach was useful in exploring the official reasoning behind the implementation of the strategies as well as its merits and limitations in Bulawayo. It also helped to under-stand residents' experiences and attitudes towards the BCC's water management strategies. Purposive sampling was used in recruiting key informants who provided detailed information about the water situation in Bulawayo. The reason for choos-ing purposive sampling is that it is a technique widely used in qualitative research to identify and select information-rich cases for the most effective use of limited resources (Patton, 1990; Palinkas et al., 2015). Therefore, interviews were carried out with key informants from the BCC who are experts in the local authority's water supply and demand management practices.

Household study participants were Bulawayo residents from suburbs that experience water scarcity. The City of Bulawayo has 156 suburbs in total (Sintummule, 2019), and I purposively selected six suburbs. Two low-density sub-urbs (Hillside & Khumalo), two medium-density suburbs (Northend & Queens Park), and two high-density suburbs (Nketa & Makokoba). Then, from a total of 184,692 household, I selected 100 households and interviewed heads of house-holds who were both men and women. Semi-structured interviews were used to in-terview residents from households because they are open, allowing new ideas to be brought up during the interview as a result of what the researcher says. According to Drever (2003: 36) "interview guides help researchers to focus a discussion on the topics at hand without constraining them to a particular format". This freedom can help the interviewer tailor their questions to the interview context/situation and the people they are interviewing. Follow-up questions during interviews are usu-ally prompted by what the interviewees say. The researcher ensured confidentiality and anonymity by referring to research participants as Respondent 1, 2, or 3 and Key informant 1, 2, or 3.

Available documents were minutes of meetings, BCC Water and Wastewater Masterplan, information brochures on water conservation, records of water-shedding schedules, and posters. These documents were analysed since they are an efficient and effective way of gathering data from manageable and practical re-sources that can support and strengthen research. Documents helped the researcher get background information and broad coverage of Bulawayo's water security situ-ation and helped contextualise the study within its setting. In addition, documents augmented information that key informants may have left out and helped trian-gulate and enhance validity. Data from interviews, minutes of meetings, the BCC Water and Wastewater Masterplan (2012), information brochures on water conser-vation, records of water-shedding schedules, and posters were presented through citations.

Water Demand Management Strategies Used by Bulawayo City Council

To address the water insecurity challenges mentioned above, the BCC has embraced water demand management and conservation strategies.

Water demand management is the adaptation and implementation of a strat-egy by a water institution or consumer to influence the water demand and usage to meet objectives such as economic efficiency, social development, social equity, environmental protection, the sustainability of water supply and service and political acceptability.

(Robinson and Eberhard 2003)

The city has therefore adopted strategies such as Water Conservation and Water Demand Management (WCWDM) to preserve water resources and meet current consumption demand by reducing water losses. This is clearly spelt out in its Water and Wastewater Plan (2012–2022). The measures implemented by the BCC include water metering and billing, water shedding, water rationing, a hosepipe ban, and the ban on purified water for construction and wastewater reuse.

Water Metering and Billing

Water metering and billing are used by water supply authorities to manage water demand and a specific consumption. The BCC is no exception, as it uses water meters to measure water usage at each property and bill customers using the meter readings. This strategy is used during times of water abundance and scarcity. Residential properties are connected to the water supply network and metered since a water meter is a requirement before the property can be connected according to the by-laws. The City of Bulawayo has a total of 134,381 occupied properties, while the total number of metered connections is 125,891, and the city is divided into 53 management meter zones for leakage control and pressure management (Key informant 7, BCC). According to the BCC's Water and Wastewater Plan (2012), the council ought to have sustainability through replacing all non-functioning water meters and ancillary infrastructure in the distribution network.

To provide customers with a correct monthly bill, meters must be maintained in good working order. This ensures that the revenue is collected with the minimum inconvenience to the utility and provides the necessary confidence to customers (Gumbo, 2004). Of the residents interviewed in all the six suburbs 90% in this study concurred that the purpose of water meters is to monitor water usage and for billing purposes. One resident also concurred with the above issue and said, "Each household has a water meter which is used to calculate monthly usage and water bills. Water metering helps the users to be cautious and not become wasteful because if you use more water, you also pay more money" (Household respondent 6, Nketa).

The BCC's water and sanitation service tariffs are regulated by the Ministry of Local Government, and costs are increased in line with usage. The city's current pricing of water service delivery to residents is guided by the rising block tariff structure that was adopted in 1992. This was implemented as a tool to curb excessive water demand in the commercial and domestic water sector with basic consumption limited to 600 l/household/day or 18 kl/month (Key informant 7, BCC). The rising block tariff structure has also accelerated efficiency and equitable fairness of distribution. Although the tariff block system is theoretically applauded for being efficient in conserving water, in practice, it penalises the poor families with large households and shared connections. Most residents interviewed have bigger families or in some instances two or three families who stay in one house which means they share water. With the block tariff if you use more water you pay more money. The advantage is for small families who use less water and, in most instances, are high-income earners. Therefore, social justice is not being fully operationalised in Bulawayo City. To cushion the poor, cross-subsidies can be used where the high-income earners subsidise the low-income earners. Or the city council can come up with pro-poor strategies for those with low incomes.

For billing purposes, the city is divided into 23 districts, which are a combination of suburbs, and these districts do not coincide with the meter zones. There are 28 meter readers, each expected to read 250 m a day in the high-density areas and 110 a day in the low-density areas and industry (Key informant 7, BCC). Ideally, meters are supposed to be read once every month, but due to staffing problems and other challenges, the meters are normally read once in two months. The city council must ensure that meters are read every month so as not to prejudice customers. Key informant 7, BCC summarised the billing process as follows:

After the readings are taken, they are captured into the database and the information technology section will produce two reports: For those readings that result in consumption that has deviated more than 5% from the average and for those accounts which did not have readings and therefore have to be estimated, these are handed over to the deviation section to check and make the necessary corrections. A test run will then be conducted in which the tariffs will be applied to the consumption for a particular month. An exceptions report that consists of accounts that have zero and very high or negative consumption will then be produced for the attention of the metering superintendent. A due date will then be put and printing of bills can now be done. Once the printing is complete, the bills are taken to the bill packing section for sorting, packing, and delivery to the consumers. Members of staff usually do the delivery on a door-to-door basis. Customers are then required to make payments at the 13 payment offices, which are distributed thus; 12 in the high-density areas and one in the city centre for industry, commerce and low-density areas. The consumers may also pay via a stop order facility and direct banking.

By-laws empower the council to cut off supply as a way of enforcing payment and charging a reconnection fee for services to continue. It was also established that most bills were based on estimates because of the high level of faulty metering and infrequent meter reading due to high staff turnover. Basing bills on estimates creates room for errors, which could result in overstating or understating the water demand in the city. Overstating water demand normally leads to overbilling consumers and destroys the customers' confidence, which is important in successfully implementing a Water Demand Management (WDM) programme. One respondent complained about the issue of estimates saying: "I once stayed for two full months without receiving water from the council, but I received a bill stating that

I should pay 30 US dollars" (Respondent 7, Khumalo). This shows the disgruntlement among residents on unfair pricing and the negative effects of estimates. To these allegations, one representative of the Water Services Sector responded:

> Firstly, residents must know that if their gates are locked and we fail to have access to the water meter we have no other option but to rely on estimates pegged at 20 mega litres a month. Secondly, the water bill also consists of other rates such as refuse collection and housing rates. It's not possible to have a zero bill after every month.
>   (Key informant 3, BCC Water Services Department)

This shows that there is poor communication between the water supply service authority and water users. There is a need for effective communication so that residents notify the council when travelling and that they are required to pay housing rates and refuse collection bills even when they are not around. By using the estimate of 20 Ml a month, the council may have either under- or overestimated water use by domestic users. Understating water demand culminates in water losses, as the water utility will not be able to account for the water that may not have been billed, and it undermines the efficient use of water. Water users who are important stakeholders lose trust with the system which is not accurate but rely on estimates. One resident also (Respondent 8, Hillside) complained that BCC officials do not come every month to collect meter readings, so in some instances, the residents are overcharged. One city council official responded and said that they are understaffed, and in some instances, there is no fuel to transport officials from the Water Services Department (Key informant 2, BCC Water Services Department).

Eighty percentage of the residents interviewed said that they were not consulted on water pricing. They professed ignorance on how water bills are calculated. Respondent 9 (Queens Park) pointed out that although the city council uses water meters for water usage, they do not understand how they charge for water usage. One resident pointed out that the BCC holds annual feedback meetings, but these are often about expenditure and not proposed budgets (Respondent 10, Hillside). One official from the Water Works Department pointed out that the city council holds budget consultative meetings and that's where issues of costs, income, and expenditure are discussed. He said that in most cases, residents shun such meetings. Some residents said the city council only sent water bills that they do not understand and need interpretation. Most residents do not even know the price of water. The council just gazettes water rates and disseminates them in newspapers and local radio stations.

The BCC's efforts to improve water infrastructure were being hindered by some residents who do not pay up their water bills, and this derails the council's efforts of providing water to the people. In some instances, their water is disconnected due to non-payment of bills. For example, in 2017, the BCC embarked on massive disconnections of water supplies to defaulting residents even though the High Court had upheld a previous ruling that such actions were illegal. The council disconnected water supplies to consumers owing US$300 or more. Those whose supplies were disconnected had to pay at least half the outstanding amount for supplies to be restored, and they had to devise a payment plan to clear balances. Section 8 of the water by-law – Statutory Instrument 164 which empowers local authorities to cut water supplies without a court order breached Section/Article 77 which guarantees the right to safe, clean, and potable water and the constitution which "imposes a duty on the state and Government institutions like councils to respect the fundamental rights and freedoms". The two sections guarantee the right to safe and clean water and compel the government to respect fundamental human rights. One Bulawayo Progressive Residents Association's (BPRA) representative pointed out that the BCC should appreciate that defaulting residents have no money to pay bills (Key informant 9, BPRA). Water must be affordable because it is a basic right. If all these groups are represented, the water pricing will seem fair so that the service provider will charge a rate that enables the BCC to continue operating while not overcharging a client and not profiteering because it is a basic right. World trends have introduced primary water or free basic water. Every first 5 kl or m^3 is for free. There is an issue of cross-subsidy because those who can pay are effectively paying for the poor. Water is both an economic and social good. By getting the first 5 m^3 of water for free, the poor will not say they do not have enough water; at least they have enough for domestic use.

During times of water scarcity, efficient technologies assist in conserving water and making little water go far. Well-managed water systems such as water meters can ensure the correct amount of water supplied and help in saving water. However, most water meters in Bulawayo are very old and must be replaced. From discussions with key BCC personnel in the water supply division and analysing internal reports by the water supply division, it was established that 21% of the meters in the city were faulty, and the council is also losing a lot of revenue through ageing meters. According to the BCC Engineer's Report presented at the Zimbabwe Investment Authority Meeting (2018), the BCC is losing an estimated US$170,669 due to ageing water meters from water meters installed before 1980. From meters installed from 2007 to now, the BCC is losing an estimated US$68,958.45. The city council has a mammoth task of replacing dysfunctional and old water meters. Because of obsolete water meters, the BCC is losing both precious water and revenue.

Prepaid Water Meters

In 2013, the BCC introduced prepaid water meters (PPWMs) after being given the green light to do so by the local government. The local authority resolved to introduce PPWMs as a water demand management strategy in Bulawayo. In line with the Integrated Urban Water Management (IUWM) approach of economic sustainability, the BCC wanted to reform its water sector to collect efficient meter readings and bills paid by residents. The revenue collected as a result of the installation of PPWMs was going to strengthen the city council's capacity to finance its water sector. The PPWM system is different from the conventional one which is used by the BCC where one pays a bill after use. One key informant pointed out that:

> Under the prepaid water system, consumers purchase vouchers linked to a credit card with a code that they feed into the meters and get credit units which are commensurate with the value. Failure to buy credit units results in one being automatically disconnected.
>
> (Key informant 3, BCC)

The BCC was projecting that the installation of PPWMs had the potential of increasing the local authority's annual revenue base to US$41,125 million annually, up from US$29,157 million as well as improving the consumer metering and billing system (BCC, 2018) the BCC had struggled to collect revenue from consumers, including residents, commercial entities, and government departments, forcing them to embark on extensive water cuts on defaulters. Also, the installation of PPWMs was the only way to ensure that residents pay for the water they consumed. The system enables the BCC to recover the money owed to it by the residents. In 2015, the BCC revealed that it was owed US$90 million in unpaid bills, and in 2018, it was owed US$93 million by residents, US$62 million by industrial and commercial entities, while parastatals and some government ministries owed it more than US$3 million (Chronicle, 5 July 2018). Former Environment, Water and Climate Minister Saviour Kasukuwere disclosed during Parliament's question-and-answer session in 2013:

> Water provision on its own requires somebody to pay for it. We agree that water is a human right, but the transmission must be paid for. So the question of prepaid water meters is meant to support councils and service providers to have enough capacity to service the communities with water.
>
> (Sunday Mail, 17 January, 2016)

The move to install PPWMs was met with mixed reactions from the residents in Bulawayo. A few of them supported the action as they felt that it was the only way to ensure residents paid for the water they consume. One official pointed out that people must pay for water just like they pay for any other services such as electricity and food. This is the case because some residents do not pay their water bills, which derails the city council's effort to improve the water system. This then affects some residents who pay the bills on time (Respondent 12, Hillside). While councils viewed the prepaid water primarily to boost revenue and improve service delivery, some sections of society felt that it was a ploy to deprive them of water, an essential commodity. Many people regard access to water (Sustainable Development Goal [SDG6]) as a fundamental human right that should easily be available to all. One female resident noted that:

> It is naïve to compare water and electricity. Water does not have any substitute, whereas electricity can be substituted with wood, solar and gas. Water

is a basic human right and once prepaid water meters are installed, the poor will not be able to access it.

(Respondent 13, Nketa, March 2014)

This resulted in protests in 2014 where civic organisations, including BPRA, Bulawayo Agenda, Restoration of Human Rights and Women of Zimbabwe Arise, lobbied Bulawayo residents against the PPWM. The BCC finally bowed to pressure from residents and halted the project following the demonstrations (shown in the picture below). The PPWMs will now be installed based on voluntary acceptance by users.

The demonstrations were a clear indication that disgruntled groups in society were not happy with the move. The majority of residents from poor communities such as Makokoba and Nketa protested to show their inability to afford the use of prepaid meters. The human right should be fulfilled even during times of climate change and people must have access to water. Using the climate justice lens eliminating inequalities in access to water and sanitation is a foundational requirement for effective climate action. Residents and grassroots organisations demonstrated against the social injustice of being denied the right to water. PPWMs must not be used to discriminate other water users, especially the poor. Using the lens of just transition adaptation technologies to water conservation must not exclude vulnerable groups in society. As a result, the BCC faced resistance when they tried to use PPWMs aimed to "enhance cost recovery and addressing the problem of non-payment" (Schwartz et al., 2017).

Water Rationing and Water Shedding

Water rationing is one of the common strategies that the BCC resorts to during water stress periods. This is meant to conserve the little water that the city is left with for that period. In recent years, Bulawayo had to introduce water restrictions from 2012 up to date. Zimbabwe's urban water management strategy uses a drought mitigation condition called 21-month rule to minimise the effects of future droughts on urban centres. In summary, the rule details that if storage at the end of a rainy season is not sufficient to sustain a city to the next rainy season, water shedding should commence immediately. The rule also gives adequate time to invest in mitigation projects to lessen the impact during water crisis and in the event that the city does not get adequate inflows in the following rainy season. The BCC advertises in local newspapers and warns residents that a water rationing scheme is going to be implemented. For example, the 2012–2013 and 2014/2015 seasons did not result in many dam inflows; for the fifth time, the city had to introduce water rationing in 2013. Similarly, the 2015/2016 rainy seasons did not result in many inflows into the supply dams and as a result, a tighter water rationing regime up to 2017 was adopted. High-density suburbs were pegged at 500 l, low-density suburbs 750 l, and cottages with meters 200 l, flats with meters 350 l/day and per household (Key informant 2). The senior resident of Hillside suburb shared his view about water rationing citing that "Water rationing schemes were and are still perhaps one of the best desperate moves set by the BCC to conserve water. But one disadvantage of them is that residents tend to suffer more during such periods of rationing" (Respondent 14, Hillside). Residents who rely on it for survival there are severely affected. Residents who have their livelihoods depend more on water such as market gardening, brick moulding, construction, fast-food restaurants are severely affected. Under the lens of just transition, climate change interventions need to secure workers' jobs and livelihoods. Therefore, the central government must provide alternative water sources or food or cash aid to households to cushion them in times of water scarcity during climate change.

Some water rationing methods also implemented by the BCC include limiting household water consumption per day and charging penalties for households that exceed the daily limit. People are not allowed to water their gardens using hose-pipes. One respondent said that these measures are reasonable as long as the water is available 24 hours (Respondent 16, Khumalo). Interest is charged if one exceeds expected consumption, and a penalty is charged for using more than the regulated consumption per household. These penalties encourage households to use water sparingly. Some residents pointed out that there are no checks to see whether these are adhered to on water rationing.

Numerous efforts to deter consumers from using hosepipes and water for brick moulding have been unsuccessful, with scores of people still engaging in these activities. The penalties imposed do not seem to deter them from the use of water for these purposes. One resident encouraged the BCC to conduct spot checks on construction projects to ensure that purified water is not utilised (Respondent 15, Nketa). Some residents do not cooperate in conserving water as they give prefer-ence to their economic activities. Some adaptation strategies employed by the BCC hinder the progress of residents' economic activities; therefore, they are unfavour-able to them. The city council and the state must provide residents with alternative water sources so as not to harm or disturb residents' economic activities. The whole narrative of these alternative water sources such as utilisation of underground wa-ter, the building of more dams, and water harvesting require climate financing.

Due to the massive water stress experienced in the past ten years (2012–2022), the city implemented a water-shedding strategy. Water shedding is whereby house-holds have their supplies completely cut out for days. The water users are informed about the water-shedding schedule and reduction of monthly water allocation per household. For example, in 2013, the BCC introduced water shedding for four days a week. Sometimes the council fails to abide by this because the water levels in reservoirs supplying some suburbs go below critical levels, rendering them un-able to pump water. Water levels in dams continue to drop due to excessive heat and evaporation. As a result, this affects the water pressure in suburbs reconnected with water supplies, especially those in high ground areas like Entumbane. The township is one of the oldest suburbs in Bulawayo and has experienced hardship in water supply channels because of its geographical location. The strategy is one of the most effective ways of conserving water, especially from the city council's perspective.

However, as a conservation strategy during water shortages, water shedding has adversely affected people's lives. The absence of water for 72 hours has affected women and the girl child at a larger scale. Sometimes women travel long distances in search of water for domestic use and gardening, which exposes them to rape, torture, and sometimes death. Furthermore, the absence of water for good sanitation compromises people's health and exposes them to diseases such as cholera. Women must not be sidelined in issues to do with climate adaptation, and their needs must be prioritised. The effectiveness of any action to scale back the impacts of climate change requires an understanding of those gender-differentiated impacts, vulnerabilities, and capacities, to handle the particular needs of women and men.

Using a climate justice lens, the residents living in high-density suburbs like Makokoba had already been disadvantaged due to colonialism (racial segregation) and contemporary dynamism of the post-colonial development state. Bulawayo residents are vulnerable to droughts and water stress. Inequalities caused by climate change are also worsened by differences in gender, class, livelihoods, and geographical location. Water shedding and rationing create threats and risk for poor households and those who rely on water for their livelihoods (cooking, car washing, construction, and brick moulding). Shortages of water increase labour for women and girls who also travel long distances to fetch water. Women and young girls end up fetching water in unprotected wells which put their health under risk and contact diseases such as dysentery, cholera, and typhoid. Without proper framing, for instance, in terms of equity and burden sharing, adaptation also promotes gender injustice.

On measures such as water rationing, water shedding, banning the use of hosepipes for watering gardens, and the ban on the use of council water for construction, residents felt the measures were punitive for water users. The BCC does not offer them adequate alternatives where they can get water as boreholes, and bowsers are not enough. They felt that the BCC is reactionary as it only resorts to punitive measures in times of drought and water scarcity. In the just transition context, there is a need for an integrated approach where all stakeholders including the water utility (BCC), the government (ZINWA), and water users are brought together to discuss the purpose of managing water efficiently.

Water Conservation Awareness Campaigns and Roadshows

Over the past years, the BCC carried out several campaigns to inform residents about the water situation, encourage them to conserve water, and support the city's efficient water use programmes. These campaigns have made it relatively manageable for the city to enforce water demand management measures such as water rationing and banning hosepipes. With the funding from the African Development Bank, the BCC carried out water conservation awareness campaigns in schools, churches, and other social gatherings and through phone messages, billboards, and printing messages on monthly water bills. One resident concurred that she was aware of posters and messages printed on monthly water bills (Respondent 16, Hillside). Although the BCC said that it usually carries water conservation awareness campaigns, 80% of residents interviewed professed ignorance about these. Residents from low-density suburbs said that the majority had not taken part in roadshows or campaigns. One key informant highlighted that they have not done much to educate people and intensify these campaigns to various institutions and forums (Key informant 3, BCC Engineering Services Department). In 2018, the BCC embarked on a door-to-door awareness campaign to encourage residents to use water sparingly. The campaign was carried out under the Bulawayo Emergency Water Augmentation Programme (BEWAP) (Chronicle, 2018). Besides education about water conservation, the teams also educated residents about basic hygiene to avoid the outbreak of diseases. The local authority was working with Dabane Trust, MEDAIR, and World Vision. The programme was carried out by community workers. Water-saving campaigns in Bulawayo also showed the highest level of stakeholder participation by cooperating with the local authority, NGOs, and residents. Most residents in the city have complied with water conservation methods. Although the BCC is making efforts to conscientise the city on water conservation methods, a lot can be done to intensify the water-saving campaigns. Dissemination of information is another pathway of just transition, and the BCC is making some major strides.

Non-revenue Water Management

In its Water and Wastewater Masterplan (2012–2022), the city council intended to replace non-functional meter and repair leaks and has started replacing them. A lack of enough funds hinders the process. On-revenue water reflects vast volumes of water lost through leaks, which are illegally connected and not invoiced to customers. (The major challenge is that most pipes have outlived their time and need replacement. Right now, the council is repairing the pipes, but it is costly to maintain those pipes. This was just a stopgap measure to avert water shortages in Bulawayo. The BCC's water utilities suffer from the substantial financial costs of treating and pumping water only to see its leak back into the ground or not being billed accurately and the lost revenues from water that could have otherwise been sold and used to maintain or further develop the network. In 2019, total water produced was 30,129,528 m^3, and water billed was 17,593,619 m^3, and non-revenue water was 41.6% (Key informant 3, Engineering Services Department). A lot of revenue was lost due to unbilled water and ageing meters in 2017. The council lost US$2,501,490,37. The BCC cited that more water was lost through leakages, faulty meters, and worn-out pipes that resulted in bursts. The high losses were attributed primarily to old infrastructure and inadequate resources to construct, manage, operate, and maintain water systems. In the short term, the city needs to rehabilitate its water distribution network, which requires enormous capital investment to meet water supply constraints. Therefore, the percentage of unaccounted for water in the system has become a measure of not only the physical condition of the system but of the system management as well.

Conclusion

Water demand management strategies employed by the BCC act as a stopgap measure during times of water scarcity. The strategy enables the city council to make the little water go far during times of crisis but does not address the fundamental causes of water shortages. For example, water rationing and shedding have ensured that the city survives the most extreme instances of potable water scarcity during the severest droughts. However, these punitive measures the council imposes during drought hit residents and industries hard and affect the already vulnerable groups. Therefore, climate change decision-makers must understand the adaptation needs and mitigation opportunities of water and sanitation systems and the risks that climate change poses to sustainable services. They must additionally align climate and water policies so that access to water is equitable, climate risks are reduced, and there is more money available for adaptation.

Also, the local authority and residents are partly to blame for the ineffectiveness of some water demand management strategies. Most residents do not pay their bills on time, or they do not pay at all. On the other hand, the council is failing to replace malfunctioning water meters and carry out extensive programmes to educate the community on water conservation strategies. All these shortcomings derail the efforts to manage the little water available in the city. To realise these rights, local authorities must be funded to improve their capacity to manage water sources sustainably and efficiently. Additionally, the BCC can use technical interventions to manage water demand. All new houses can be built in such a way that water is collected from roofs and stored in tanks. Therefore, to protect the susceptibility of urban dwellers in arid and semi-arid regions such as Bulawayo from climatic vagaries and enhance resilience to climate change, optimising rainwater harvesting and storage techniques has become the most ideal alternative climate change adaptation strategy. When surface runoff is trapped into reservoirs, it helps manage droughts.

References

Alexander, C., 2019. Cape Town's 'day zero' water crisis, one year later. https://www.citylab.com/environment/2019/04/cape-town-water-conservationsouthafricadrought/587011/ (Accessed 25 November 2019).

Bhatasara, S. and Nyamwanza, A.M., 2022. Climate injustice and the role of climate justice movements in Africa: the case of Zimbabwe. In *Africa's Radicalisms and Conservatisms* edited by Edwin Etieyibo, Obvious Katsaura, and Muchaparara Musemwa, (pp. 187–209). Boston, MA: Brill.

Beukman, R., Reeler, R. and WWF South Africa, 2021. A just transition in the water sector: policy brief for the Presidential Climate Commission. https://www.climatecommission.org.za/events/dialogue-on-water-security.

Bulawayo City Council, 2012. Water and wastewater plan, IMIESA, Vol 37 (11), https://hdl.handle.net/10520/EJC127632.

Chilunjika, A., and Zhou, G., 2018. A peep into the sources of policy implementation inertia in Africa: The case of the Matebeleland Zambezi Water Project (MZWP) in Zimbabwe. *Asian Journal of Empirical Research 3* (4), pp 447–463.

Drever, E. 2003. *Using semi-structured interviews in small-scale research: A teacher's guide*. Glasgow: the SCRE Centre, University of Glasgow.

FAO, 2015. Climate change and food security: risks and responses. https://www.fao.org/3/i5188e/I5188E.pdf.

Gumbo, B., 2004. The status of water demand management in selected cities of southern Africa. *Physics and Chemistry of the Earth, Parts A/B/C, 29*(15–18), pp. 1225–1231.

ILO, (2019). Climate change and financing a just transition. https://www.ilo.org/empent/areas/social-finance/WCMS_825124/lang--en/index.htm.

Mabiza, C.C., 2013. *Integrated Water Resources Management, Institutions and Livelihoods under Stress: Bottom-up Perspectives from Zimbabwe; UNESCO-IHE PhD Thesis.* Milton Park, Abingdon-on-Thames, Oxfordshire: CRC Press.

Makwara, E.C., 2012. Water woes in Zimbabwe's urban areas in the middist of plenty: 2000-present. *European Journal of Sustainable Development, 1*(2), pp. 151–151.

Mapuva, J. and Takabika, T. 2020. Urban local authorities in Zimbabwe and the new constitution. International Journal of Peace and Development Studies, 11(1), 1–8.

Markkanen, S. and Anger-Kraavi, A., 2019. Social impacts of climate change mitigation policies and their implications for inequality. *Climate Policy, 19*(7), pp. 827–844.

Maviza, A., Grab, S. and Engelbrecht, F., 2022. Twentieth century precipitation trends in the upper Mzingwane sub-catchment of the northern Limpopo basin, Zimbabwe. *Theoretical and Applied Climatology, 149*(1–2), pp. 309–325.

Mkandla, N., Van der Zaag, P., & Sibanda, P. 2005. Bulawayo water supplies: Sustainable alternatives for the next decade. *Physics and Chemistry of the Earth*, Parts A/B/C, 30(11–16), 935–942. Amsterdam: Elsevier.

Moyo, B., Madamombe, E., Love, D. and WaterNet, P.O., 2005, November. A model for reservoir yield under climate change scenarios for the water-stressed City of Bulawayo, Zimbabwe. In *Abstract Volume, 6th WaterNet/WARFSA/GWP-SA Symposium, Swaziland* (p. 38).

Musemwa, M., 2014. *Water, History, and Politics in Zimbabwe: Bulawayo's Struggles with the Environment, 1894–2008.* Trenton, NJ: Africa World Press.

Newell, P. and Mulvaney, D., 2013. The political economy of the 'just transition'. *The Geographical Journal, 179*(2), pp. 132–140.

Noemdoe, S., Jonker, L. and Swatuk, L.A., 2006. Perceptions of water scarcity: the case of Genadendal and outstations. *Physics and Chemistry of the Earth, Parts A/B/C, 31*(15–16), pp. 771–778.

Palinkas, L.A., Horwitz, S.M., Green, C.A., Wisdom, J.P., Duan, N. and Hoagwood, K., 2015. Purposeful sampling for qualitative data collection and analysis in mixed method implementation research. *Administration and Policy in Mental Health and Mental Health Services Research, 42*, pp. 533–Patton, Q. 1990. Qualitative evaluation and research methods, Newburypark, CA: Sage Publications.

Porter, L., Rickards, L., Verlie, B., Bosomworth, K., Moloney, S., Lay, B., Latham, B., Anguelovski, I. and Pellow, D., 2020. Climate justice in a climate changed world. *Planning Theory & Practice, 21*(2), pp. 293–321.

Robinson, P. and Eberhard, R, 2003. Guidelines for the development of national water policies and strategies to support integrated water resources management, (Draft) SADC, Water Sector Coordination Unit, Gaborone.

Rutger W.F et al. 2019. 17 Countries, Home to OneQuarter of the World's Population, Face Extremely High Water Stress, World Resources Institute, https://www.wri.org

Schwartz, K., Tutusaus, M. and Savelli, E., 2017. Water for the urban poor: balancing financial and social objectives through service differentiation in the Kenyan water sector. Utilities Policy, 48, pp. 22–31.

Sibanda, T., Nonner, J. C., & Uhlenbrook, S. 2009. Comparison of groundwater recharge estimation methods for the semi-arid Nyamandhlovu area, Zimbabwe. Hydrogeology Journal, 17(6), 1427.

Sibanda, K. 2018. The Politics of Water Crisis in the City of Bulawayo, Proceedings of the SADC International Conference on Postgraduate Research for Sustainable Development, School of Graduate Studies University of Botswana file:///D:/Downloads/PROCEED-INGS%20-%20SADC%20ICPRSD2018.pdf

Sintummule, N.I. and Mkumbuzi, S.H., 2019. Participation in community-based solid waste management in Nkulumane suburb, Bulawayo, Zimbabwe. Resources, 8(1), p.30.

United Nations (UN), 2013. What is Water Security? Infographic https://www.unwater.org/publications/what-water-security-infographic

United Nations (UN), 2018. World water development report. https://www.un.org/waterfor-lifedecade/scarcity.shtml.

United Nations (UN), 2022. World water development report. https://www.unwater.org/publications/un-world-water-development-report-2022/#.

USAID, 2020. Zimbabwe climate risk profile. https://www.climatelinks.org/sites/default/files/asset/document/2020_USAID_ATLAS_CRP-Zimbabwe.pdf.

UNICEF, 2019. The state of WASH financing in Eastern and Southern Africa, Zimbabwe Country Level Assessment https://www.unicef.org/esa/sites/unicef.org.esa/files/2019-10/UNICEF-Zimbabwe-2019-WASH-Financing-Assessment.pdf

Zimstat,2022.https://www.zimstat.co.zw/wpcontent/uploads/Demography/Census/2022_PHC_Report_27012023_Final.pdf

5 Climate Finance, Public–Private Partnerships and Climate Injustice in Lesotho

Thapelo Ramalefane and Philani Moyo

Introduction

Climate finance is public or private funds channelled through government or recognised non-governmental entities for climate action, specifically mitigation, adaptation and resilience activities (Mahat *et al.* 2019; Warren 2019). Climate-financing protocols emerged from the international climate governance system under the United Nations Framework Convention on Climate Change (UNFCCC) as a deliberate effort to ensure that climate actions are well resourced and mainstreamed into national development programmes as part of efforts towards climate-resilient development (Barrett 2014). This climate-resilient development agenda predominantly involves radical changes in country development pathways, industrial production processes and systems, material consumption choices for the citizenry and changes in the human socio-environmental nexus. These changes require substantial financing as the macro economy and broader society undergo structural adjustments and behavioural shifts towards reduced carbon emissions and climate-resilient development (Sithole and Murewi 2009). This means the primary rationale of climate financing is about assisting countries to transition towards low-carbon development (through mitigation) while simultaneously providing resources for different economic, environmental and social actors to adapt and build resilience to climate impacts.

There are various perspectives that ideologically inform and are used to explain the rationale of climate finance. In this chapter, we argue that climate finance is underpinned by the notion that it is only fair and just for well-resourced Global North (the North) countries to financially support mitigation, adaptation and resilience activities in the Global South (the South) (Vanderheiden 2015). This is because the North has historical primary causal responsibility over the current state of the climate emergency and must thus contribute more in responding to it. Their predominant financial and technical contribution is also a form of compensation to the South because the raging climate impacts in the South are fundamentally the consequence of capitalism and neoliberal globalisation driven by the North (Khan *et al.* 2020). As Khan adds, this compensation finance is also a form of the 'polluter-pays principle' that is a standard practice in international environmental and climate governance.

DOI: 10.4324/9781003397120-7

Climate finance is also directly tied to climate justice. In its elementary form, climate justice advocates argue that since climate impacts are deepening vulnerability, poverty and inequality of the already poor and marginalised, then these people are owed compensation by those who primarily caused the climate emergency (Porter *et al.* 2020). At the centre of this compensation argument is the need to improve the well-being, human development and protect environmental rights of the poor. To achieve this, those responsible for the climate emergency must clean up and compensate for the damages caused by their historical climate system destructive practices. In this sense, climate finance mediates climate justice and compensatory justice (Schlosberg and Collins 2014). Further, using procedural and distributive justice arguments, climate justice takes into account local climate impact experiences of communities, drivers of broader inequity and the extent of local people's participation in climate action (Schlosberg and Collins 2014) to demonstrate how they continue to suffer the most under the climate emergency, hence the justification for their compensation. Within this context, this chapter examines a climate finance public–private partnership in Lesotho, with a specific focus on its modalities, ideological and bureaucratic politics as well as inefficiencies at a local community level. It is informed by and analyses the lived experiences of Thuathe and Khubetsoana smallholder citrus farmers in peri-urban Berea (Lesotho) who are beneficiaries of climate finance raised and distributed through a public–private partnership. In doing so, this chapter answers three specific questions: (1) What are the modalities of public–private climate financing for smallholder citrus farmers in Lesotho? (2) To what extent is this public–private partnership enhancing smallholder farmers' adaptation and resilience as pathways towards distributive and procedural justice. (3) What are the obstacles of the climate finance public–private partnership in mediating climate justice?

'Broken' Climate Finance Promises and Illusions of Climate-Resilient Development

Climate finance sources are multiple and diverse in their origins, orientation, focus and operational procedures. In the main, most of the funds mainly originate from Global North governments, multilateral and bilateral sources (Barrett 2014) as well as private sector entities. This diversity of sources means competing political, ideological, environmental, economic and social interests are central players in the international climate-financing field. With some overlap, they all pursue different aims, goals and targets for their own geostrategic means and ends. This explains why different actors attempt to address climate impacts through funding a programme or project aligned to their geostrategic agenda.

The biggest and well-known international climate-financing mechanism operates under the direct and indirect auspices of the UNFCCC, Paris Agreement on Climate Change and associated global implementing agencies and national implementing entities. At a global level, some of the prominent implementing agencies include the Global Environment Facility (GEF), Green Climate Fund (GCF), Adaptation Fund (AF), Least Developed Countries Fund (LDCF) and the Special Climate

Change Fund (SCCF). The overarching idea is to use these multilateral and bilateral channels to raise, manage and distribute climate finance especially to the South (Huang and Wang 2014) using a globally synthesised financial mechanism that can be monitored and its performance evaluated. Although figures vary, the initial non-binding agreement under this financial mechanism was that northern countries will finance climate action in the South to the tune of US$100 billion per year by 2020 (Brown *et al.* 2010; Hallegatte *et al.* 2016; Steckel *et al.* 2016; Stalley 2018). This figure, US$100 billion (by the end of 2020), subsequently became binding as from 2016 under the Paris Agreement. In addition, developed countries would augment developing countries' annual budgets directly, thereby also making significant contributions to the latter's ability to adapt and build resilience to climate impacts (Kameyama *et al.* 2016; Yamineva 2016). The agreement was that these funds will be sourced from different governments (at all scales), multilateral, bilateral and private sources. Although hundreds of millions of dollars were, and are still being, transferred from the North to the South for mitigation, adaptation and resilience, the target of US$100 billion was, for a phalanx of reasons, not achieved. In simple terms, the US$100 billion target was not met. Failure to clearly outline whether the US$100 billion was only new additional funding, excluding previously received climate financing as well as inability to specify the exact sources counting to this total (Brown *et al.* 2010; Steckel *et al.* 2016) are some of the reasons why this target was never met. While some might argue that the US$100 billion was perhaps an ambitious target hence failure to achieve it by the end of 2020, it remains a fact that many Global North countries were (and are) simply not committed to honouring their climate finance pledges and targets under the Paris Agreement. For a variety of reasons, they are also not under any local political pressure or geostrategic push to actively meet international climate-financing targets.

The genesis of this uneasy funding relationship (between North and South) is acknowledgement that climate change is predominantly a function of North countries' historical carbon-intensive developmental processes more than it is of the South (Chinoswky *et al.* 2010). Due to this variance in greenhouse gas (GHG) emissions that cause global warming, the accepted global position within a climate justice perspective is that North countries must provide the bulk of financial and material resources for mitigation, adaptation and resilience building. This climate justice-oriented reasoning thus underscores the current North to South climate finance flows (Urpelainen 2012). Using these climate justice lenses, our argument is that many Global South countries face climate mitigation, adaptation and resilience constraints due to historical and ongoing capitalist and neoliberal production and reproduction processes that are responsible for the climate emergency. The international climate finance regimen must therefore remain alive to this historical and ongoing reality. This reality is also the reason why it makes sense to invoke the UNFCCC's principle of 'common but differentiated responsibility (CBDR)' in justifying why it is the primary responsibility of North countries to avail more climate finance to South countries (Grasso 2009). Notably, the CBDR principle has its origins in developing countries as they lobbied and negotiated for climate finance from developed countries (Stalley 2018), and because of this, it can be argued that they played an important role in setting the agenda and pathway of the global climate finance system with elements of climate justice. However, this does not mean that North-dominated geopolitics and climate politics were not, or are not, central in determining the direction and power dynamics about which country gets how much money, when and for what mitigation and adaptation interventions.

Although North countries continue to dominate climate-financing economics and politics, countries of the South have, over the years, coalesced around a climate justice agenda with a strong voice amplified by environmental movements and civil society. The climate justice agenda and action "acknowledges that because the world's richest countries have contributed most to the problem, they have a greater obligation to take action" (Adams and Luchsinger 2009: ix) as part of their historical responsibility (Fisher 2015). Their financing responsibilities are not only affirmation of CBDR principles but also reflect global mutual responsibility towards ensuring that South countries "have equal opportunities to develop" (Adams and Luchsinger 2009: xii). This means climate justice is not only a matter of equity and human rights (Fisher 2015) but also an ethical, economic and social (Adams and Luchsinger 2009: ix) imperative. Informed by and anchored on this climate justice agenda, South countries argue that they are entitled to finance from North countries not only for mitigation and adaptation but capacity building and technology transfer as well (Huq *et al.* 2011; Corsatea *et al.* 2014). Interestingly, many North countries, except the United States of America under the administration of former President Donald Trump (2016–2020), have no qualms with this climate justice ideological and funding agenda. While endorsing climate justice reasoning and action, they, however, argue that the fund-raising base must be expanded and be more innovative. The suggested additional sources include national government budget votes, climate (or carbon) tax, green incentive mechanisms, national private–public partnerships and an enabling environment for market forces to raise climate capital (Aglietta *et al.* 2015; Pauw and Bendandi 2016). However, from a climate justice perspective, these North countries suggested alternative sources of funding are problematic for so many reasons. By pretending to expand the climate finance matrix, what is actually being proposed is that a huge part of the fund-raising responsibility be shifted to governments of the South so that they raise part of the money from local taxes, budgets and financial markets. This classical neoliberal reasoning seeks to further tax and extract more money from poor South countries for climate action. It is an untenable proposal that seeks to transfer this unjust burden to poor countries, many of whom are in a debt trap and perennially saddled with budget deficits.

Further, the North to South climate-financing flow model is murky with a bias towards private capital, in other words, big business. There is evidence that a big chunk of the funds are used as capital investments in mitigation efforts, dominated by private investors, more than in grassroots adaptation endeavours (Rojas-Downing *et al.* 2017). This high participation of private investors in mitigation programmes and projects, for example, in renewable energy, is simply because there is a reasonable chance of short- to medium-term investment returns (profits) in that sector compared to channelling funds towards adaptation and resilience-building programmes for poor and vulnerable smallholder farmers in rural communities. In addition, many mitigation activities tend to attract private investment because they fit easier in their business models than adaptation initiatives (Perkins and Nachmany 2019). This also partly explains why mitigation funding predominates the international climate finance landscape on a 'business-as-usual basis' (Rojas-Downing *et al.* 2017) because, as we know, the interests of many private investors are not always for the public good but profits. This imbalance in the distribution of financing, between mitigation and adaptation activities, is a matter that parties to the Paris Agreement must address in their regular Conference of the Parties (COP) meetings. The current practice where the majority of adaptation interventions are financed through public funds while the private sector predominantly uses its resources for mitigation activities needs to change. There needs to be an equal balance in dispensing private and public finance for mitigation and adaptation activities.

Although the international climate-financing mechanism remains the epicentre of mitigation, adaptation and resilience funding, over the last three decades, there has been renewed emphasis on public–private funding partnerships for climate action. This reinvigoration is because, until recently, there has been over-focus on North to South climate finance flows and a lackadaisical approach towards other climate-financing alternatives such as public–private funding (Bowen *et al.* 2017). Although the concept of public–private partnerships remains contested, it is generally accepted that they are "a long-term contractual arrangement between the public and private sectors where mutual benefits are sought..." (Perkins and Nachmany 2019). In some cases, this partnership may require that

> the private partners deliver the service in such a manner that the service delivery objectives are aligned with the profit objectives of the private partners and where the effectiveness of the alignment depends on a sufficient transfer of risk to the public partner.
>
> (Harvey *et al.* 2019)

Unsurprisingly, the profit agenda in the partnership model is clear. Notwithstanding this, when it became clear, especially in the aftermath of the Copenhagen Accord of 2010, that the traditional government, bilateral and multilateral finance sources were failing to meet annual climate finance targets (Gampfer *et al.* 2014), mobilisation for public–private partnerships and alternative funding sources gained currency. Subsequently, diversification of finance instruments to include public–private partnerships as well as alternative sources of funding (Bracking 2015; Boissinot *et al.* 2016; Kameyama *et al.* 2016) boomed and is now an established instrument. These partnerships have gained prominence, for example, as market-based instruments that avail finance in the climate-smart technology and energy sector (Kameyama *et al.* 2016) as well as in climate-smart agriculture adaptation and resilience building in countries such as Lesotho.

Climate Finance in Lesotho: A Primer

Climate action financing in Lesotho is largely through government, bilateral, multi-lateral, public–private partnerships and, to some extent, development aid. Over the years, and as part of its human development mandate, the government of Lesotho (GoL) has been making budgetary allocations to support general agriculture production and climate-resilient agrarian development. Different line ministries have received government financing to undertake these activities. One of the leading implementing government departments in that regard is the Ministry of Agriculture and Food Security (MAFS). Through the MAFS, and other ministries, the GoL thus makes direct financial support to agriculture in general and climate resilience building. In addition, farmers receive extension support services, advisory services and training through various GoL ministries and government departments. However, it is well documented that the GoL has not invested adequate climate finance towards mitigation, adaptation and resilience building in the agriculture sector. This is simply because it doesn't have sufficient financial resources for this social investment. As a result, it heavily relies on bilateral, multilateral and public–private partnerships for this purpose. Through these channels, by 2013, Lesotho had received US$92.1 million multilateral funding from the United Nations Development Programme for adaptation programmes (Lesotho Meteorological Services [LMS] 2013). In addition, by the end of 2019, the GoL had channelled US$300 million towards 17 climate change-related projects co-financed with international development agencies. Further, it is worth noting that the GoL also receives budgetary support from northern governments such as Canada, Wales and Ireland for climate programming (Ministry of Energy, Meteorology and Water Affairs 2013). This means multilateral and bilateral financing forms a sizeable component of the country's climate action programmes. Some of these streams of bilateral funding are specifically focused on climate change information sharing, policy formulation, capacity building and adaptation interventions (LMS 2013), all of which aim to build climate resilience.

One of the biggest multilateral institutions active in climate actions in Lesotho is the World Bank (the Bank). Over the years, the Bank has provided extensive loan and grant support to the GoL through different rural development programmes like the Smallholder Agricultural Development Project (SADP), the Horticultural Productivity and Trade Development Project (HPTD), Rural Financial Intermediation Programme (RUFIP) and the Private Sector Competitiveness and Economic Diversification Project (PSCEDP). One of the PSCEDP's primary aims is to alleviate food insecurity in highly impoverished areas of Lesotho (Ministry of Trade, Cooperatives, Industry and Marketing 2017). Another PSCEDP priority is to provide smallholder farmers with finances and marketing skills for their produce/products with the aim of increasing their competitiveness and high-end market penetration (Ministry of Trade, Cooperatives, Industry and Marketing 2017). As we demonstrate further below, our case-study smallholder farmers in Thuathe Orchard benefitted from these incentives by the PSCEDP. Further, the PSCEDP together with Lesotho National Development Corporation also engaged South African companies to advance horticultural businesses in Lesotho. In November 2018, the Lesotho National Development Corporation signed a M150 million deal with a South African company Stargrow Group (Pty) Ltd to commercialise and upscale the deciduous fruit industry in the country. This partnership was financially and technically supported by the Bank and run under the Ministry of Trade Cooperatives Industry and Marketing. As discussed further below, Thuathe farmers were one of the beneficiaries of this funding, new technologies, farming skills and technical support. The aim was to upscale their production and to enhance the quality of fruit produced in order to meet the targets and standards of domestic and regional markets.

In addition, the private sector has, through various market-related instruments, been a player in climate action in Lesotho. For example, through the US$4.2 million prearranged by the Bank through the Country's Partnership Framework (CPF), the private sector has access to this funding for adaptation in areas such as climate-smart commercialisation of agriculture and climate-resilient water resource development (Hallegatte *et al.* 2016). The Bank is the 'midwife' of this private sector climate-financing mechanism and thus a major player in Lesotho's climate action agenda. For this reason, and given the Bank's history in human development affairs across the country, in this chapter, we further explore how the partnership of the Bank and GoL is driving climate financing for adaptation and resilience among smallholder farmers in the country. As alluded to above, our interest is in examining the extent to which this partnership is a pathway towards reducing farmers' vulnerability to climate impacts, ameliorate socio-economic inequity and build resilience on the road to climate justice.

Case Study: Khubetsoana and Thuathe Farmers

Our analysis is informed by the lived and production experiences of citrus fruit farmers of Khubetsoana and Thuathe in peri-urban Berea, Lesotho. In both farming communities, purposive and snowball sampling was used to select a grand total of 30 farmers with active participation and production of citrus fruits being the primary criteria. An additional seven key informants, directly involved in farmer support and capacitation, were purposively selected from different government departments and the private sector. The respondent farmers traditionally focused on maize and other cereals production. However, as a livelihood diversification strategy and in response to dwindling maize output due to climate change impacts, they branched out into producing citrus fruits that are tolerant of the climatic conditions of Lesotho. These include peaches (*prunuspersica*), apples (*pyrusmalus*), apricots (*prunusarmenica*) and quinces (Ministry of Energy, Meteorology and Water Affairs 2013) which grow and mature well in the local drought and heat-prone environment. In an effort to enhance these farmers' climate adaptation and resilience strategies, the GoL and non-state actors, primarily the Bank through PSCEDP, have provided technical, technological and financial support for their production activities.

Climate Finance for Climate Justice in Thuathe and Khubetsoana

Public finance, that is GoL funding, and agricultural technical support services are central pillars of farming activities in our case-study communities of Thuathe and Khubetsoana. Up until the late 1990s, the smallholder farmers of Thuathe and Khubetsoana were mostly specialising in producing subsistence cereals, mainly maize. The farmers' own historical memory and recollection confirmed that this cereal production output was adequate to meet household food needs till the subsequent harvesting season. This suggests households were relatively food secure at the time. However, due to the climate crisis impacts (high temperatures, low rainfall and recurrent droughts), production levels plummeted with direct implications on household self-provisioning and food self-sufficiency. In response, the farmers actively adapted through switching to drought-tolerant citrus fruit farming ably encouraged and supported by the GoL's Department of Forestry. Switching to citrus fruit farming was also an on-farm diversification strategy towards eventual commercialisation. Since this switch and diversification, the majority of Thuathe and Khubetsoana smallholder farmers confirmed being beneficiaries of GoL climate adaptation and resilience finance. They have also received on-farm production technical advice and support from the Department of Forestry and the MAFS. This was reiterated by a horticulture officer at the MAFS who indicated that the ministry is involved in the Thuathe farming community because climate resilience interventions in the fruit farming sector are part of their core operational mandate. This governmental obligation includes the formulation and implementation of advisory and other technical services to farmers. In addition, the MAFS and other GoL departments have promoted and achieved relative success in disseminating climate-smart agriculture techniques that cause little disturbance to the soil, preserving water and soil nutrients. One of these is the *likoti* system (single hole for each plant) and mulching, both of which are timeous and relevant in the context of a changed climate. Further, there is also evidence of the GoL offering training and capacity building to Thuathe and Khubetsoana farmers through different agricultural institutions. For example, the Department of Agricultural Research generates, adapts and transfers climate-smart technologies and knowledge to smallholder farmers through the Agricultural Research Centres (ARC) in the districts. Moreover, these ARCs also carry out climate resilience research in collaboration with other public institutions like the National University of Lesotho, Lesotho Agricultural College (LAC) with the primary aim of enhancing the adaptation strategies of the smallholder citrus farmers.

In all these climate actions, GoL ministries and departments place emphasis on community consultation, participation and engagement for locally relevant climate action. This suggests that the government's climate-financing strategy has elements of procedural justice as it takes into account the local, social and environmental context as well the views and voices of local farmers. These elements of procedural justice further point to the government's attempt to factor in equity and justice in its climate-financing decision-making and implementation. This is a notable step by the GoL because if climate finance is universally disbursed without paying attention to pre-existing and deepening inequalities and inequities, this will jeopardise the path towards climate justice. Further, the procedural justice steps of the GoL call on us to reflect on the idea and practice of 'climate finance readiness'. Climate finance readiness is a term increasingly used to "refer to the processes at regional, national and local levels through which developing countries get 'ready' to access, allocate, distribute, and make use of financial resources for climate change action…" (The Nature Conservancy Climate Change Program 2012: 1). In the context of our Lesotho case-study communities, climate finance readiness is therefore about the GoL's aptitude and preparedness for equitable distribution in cognisance of the local needs of beneficiary smallholder farmers. Indeed, for procedural and distributive justice to be achieved, underlying socio-economic and political issues of the recipient communities (Colenbrander *et al.* 2018) and their intersection with climate impacts must be at the epicentre of distribution and utilisation of climate finance.

Notwithstanding the GoL's positive climate actions that are a step towards climate justice, it remains a fact that its interventions are few, fragmented and inadequate to drive the country's smallholder farmers to full and effective resilience. One of the reasons for this, among many others, is that the GoL departments are failing to raise adequate climate finance from national, continental and international sources. It's LMS which is the focal and responsible national entity for raising mitigation and adaptation finance is not equal to the task, or in other words, 'not fit for purpose'. With a relatively small human resource base, limited climate finance expertise and technical capacity, the LMS does not have the capability to collect all required technical information, do requisite risk assessments and preparation of detailed proposals for submission to funders (Gwimbi 2017). This incapacity, according to one of the key informants, means the LMS (on behalf of the GoL) routinely fails to meet pre-set climate action funding criteria. The country thus loses out on many funding opportunities with direct implications on its ability to implement interventions to address inequitable climate impacts at the grassroots level. Consequently, the GoL heavily relies on multilateral and bilateral support for interventions in farming communities that include Thuathe and Khubetsoana. Prior to the flagship PSCEDP, in 2009, the GoL partnered the Food and Agriculture Organization (FAO) to capacitate and distribute climate-smart technologies as a step towards on-farm diversification. While not all farmers received training about food handling and preservation under this intervention, there is evidence that a quarter of our respondent farmers were part of the programme. On the downside, the majority who didn't participate nor benefit are an indication of the targeting limitations of this resilience-building initiative.

These shortcomings of the GoL's climate action programming opened an avenue for the involvement of one of the neoliberal project's global pioneers and drivers, the Bank. As alluded to above, the Bank, with all its discredited neoliberal reasoning and programming that has wreaked havoc in the South, is at the epicentre of financing climate actions in Thuathe and Khubetsoana. Besides PSCEDP, some of the climate-smart agriculture programmes in the study areas that were funded by the Bank include the HPTD and the SADP. Even though these were largely community facilitation programmes (capacity building and information sharing) designed to train smallholder cereal producers to transition to commercial fruit farming, they also occasionally transferred climate-smart technologies and infrastructure to farmers especially those in Thuathe.

The Bank's flagship private–public PSCEDP pioneered new ground for private sector investment in the fruit farming sector. This partnership primarily concentrated on citrus fruit farmers in areas surrounding Maseru, Berea and Leribe. In typical neoliberal fashion, the PSCEDP's starting point was to push Thuathe smallholder farmers to intensify and increase production on every inch of farm land without regard to the implications of this on maintaining soil nutrients and quality, especially within a changed climate context. In further pursuit of the neoliberal agenda, farmers were also enticed to increase the quantity and improve quality of their fruits in order to satisfy local, regional and international market tastes and demand (Ministry of Trade, Cooperatives, Industry and Marketing 2017). On-farm in Thuathe, Global Good Agricultural Practices officers extensively inspected agro-chemical use, residue storage, produce labelling, day-to-day oversight, water use and contamination, crop management to ensure compliance with global market demands. It is striking, but not surprising, that agro-chemical use for production was encouraged because this is a standard farming procedure in contexts where the profit motive is the overriding imperative over environmental conservation and sustainability. The fact that these pesticides and herbicides became central in production processes among the smallholder farmers demonstrates how the capital accumulation interests of the private partners subsequently became the driving factor in the model instead of the initial objective of enhancing adaptation and resilience strategies of the farmers.

The PSCEDP private–public partnership model extended beyond farm technical assistance, supply chain management (e.g., bookkeeping) and investment promotion (e.g., development of marketing tools) to include substantial farmer financial support, specifically in Thuathe. This financing was directed towards upgrading of farm standards to meet requirements of international formal markets, assessment of demand and buyer sourcing strategies (Ministry of Trade, Cooperatives, Industry and Marketing 2017). While the Bank and GoL argue that this led to the emergence of more agricultural entrepreneurs, we argue further below that this is not the case because the partnership model failed to address the underlying vulnerabilities of the farmers and neither did it enhance their productivity for resilient, sustainable self-sufficiency. Instead, it is clear that the increased involvement of the private sector, on commercial terms not climate finance grants, improved their profit margins through increased selling of farm inputs like seeds, pesticides, herbicides, fertiliser and other inputs. Some of the main private sector beneficiaries of this in Thuathe and Khubetsoana are the major farming suppliers that include Letlotlo Farm Feeds, Bhuti Farm Products and Agriproducts. For this, and other reasons, the partnership model was not that 'climate-smart' after all as it failed to enhance the adaptation and resilience of farmers but line the profit margins of the private sector.

Another major limitation of the PSCEDP was that financial support for farmers was time limited within the project life cycle. At the time of data collection, the financing windows spanned three to seven years (Ministry of Trade, Cooperatives, Industry and Marketing 2017). Other bank-linked financing support initiatives such as SADP were for even fewer years. The implications of this are obvious: without sustained long-term funding, there was no reasonable prospect for citrus trees to be planted, nurtured and mature to full productivity. These fruit trees require years to reach their minimum production capacity. Hence, by the end of the PSCEDP project life cycle, farmers were left to continue independently before their orchards were fully productive. While this did not collapse their farming activities, it serves to demonstrate that shifting to climate-resilient citrus production requires time, money and climate-smart skills to realise its full potential. Relatedly, it shows the limitations of the private–public model where farmers are seen and treated as beneficiaries within a fixed time period instead of being equal partners that are active social agents seeking to build resilience over a sustained timeframe.

Voices and Actions of Non-state Actors: NGOs and the People

As is typical of many Southern African countries wherein government spearheaded climate programming fails, pro-poor NGOs stepped in to fill gaps of the public–private partnership model. As confirmed by the farmers, international and local NGOs are funding different adaptation and resilience-building projects in the study communities. These include World Vision, Lesotho Red Cross Society and Lesotho Council of NGOs. Lesotho Red Cross Society contributed over 3,000 peach trees as a humanitarian response to the 2015 drought, while World Vision trained farmers in conservation agriculture donated fruit seedlings and farming equipment in Khubetsoana. In addition, the Rural Self-Help Development Association (RSDA) collaborates with the Department of Forestry to disseminate new climate-smart fruit farming techniques. A notable feature of the NGOs' intervention is the promotion of conservation agriculture and climate-smart fruit farming techniques, both of which aim to enhance farmers' adaptation and resilience. Informed by these, it can be reasonably argued that the NGOs were/are indeed driven by the desire to economically uplift the smallholder farmers through enhancing self-initiated farming activities. This recognition, and enhancement, of their social agency speaks to observing procedural justice as their voices and locally relevant climate actions are advanced.

Nevertheless, even though these NGO technical and equipment interventions in the study communities slightly improved adaptation and resilience strategies, they did not, as we argue further below, comprehensively address the existential challenges they face in their daily lives. Another major weakness of these NGO interventions is that they are project based (hence limited project life cycle) and fragmented with different non-state actors working in isolation guided by their programming timeline. Hence, in as much as World Vision, Lesotho Red Cross Society, RSDA and Lesotho Council of NGOs have partnered communities in their attempts towards climate justice, there was no evidence of joint implementation suggesting they work in silos.

The limitations of the project paradigm of the private–public partnerships were also highlighted by the farmers in both Thuathe and Khubetsoana. A recurring observation was that some of their preferred crop choices were declined by PSCEDP-linked private partners and GoL officials on the basis that they are not investable since they do not fall within the funding scope of the project. Farmers also indicated that processes of accessing support were complicated by too much paperwork, administrative hurdles and language barrier as all documents are in English. Further discussions with farmers demonstrated their desire to acquire post-harvest technologies which would enhance their processing, storage and retail of produce. For instance, over 60% of farmers in both Thuathe and Khubetsoana expressed their desire to be able to afford machinery in order to process their excess produce into other valuable products like dried fruit and juices. This indicates that while financing has focused on other important on-farm aspects of fruit farming, it is clear that there is dearth of capital to finance post-harvest technologies that have potential to build resilience through onward income generation for household self-provisioning.

In addition, in a discussion about the adaptation challenges they face, 80% of the farmers attributed their problems to lack of access to up-to-date climate-resilient agriculture information held by government institutions. The complex nature of the climate emergency requires up-to-date information sharing, awareness and resources on-farm in order to inform adaptation actions in real time. It is therefore concerning that climate education, information dissemination and access to resources remain a challenge due to administrative red tape of government systems, poor government inter-departmental coordination and inconsistency in information dissemination. Consequently, many smallholder farmers are not aware of financial, human and technical resources that are available for their climate actions. Due to this, many farmers are not informed and empowered enough to take advantage of the 'hard and soft skills' and resources availed through some of the positive elements of the public–private partnerships. It also follows that these discrepancies in climate action are not only a disservice to the farmers, but they exacerbate their current vulnerabilities, deepen inequality and inequity. This suggests that more targeted government-led engagement is required in the farming communities so as to set them on a path of climate-resilient agriculture production informed by current and locally relevant information and resources.

Conclusion

The climate emergency in Lesotho is exacerbating pre-existing vulnerabilities, inequalities and inequities as it is disproportionately affecting the most vulnerable and marginalised community of smallholder farmers. Although they contributed the least to causing the climate emergency and despite having fewer material and technical capacity to build sustainable resilience, these farmers are active agents that are implementing various adaptation strategies. Through diversifying from traditional cereal production to drought-tolerant citrus fruit farming, the farmers are attempting to build resilience but with very limited production success. A combination of ravaging climate impacts, limited inputs, insufficient infrastructure, inadequate citrus production expertise and lack of technical know-how partly explains failure of this farmer-initiated resilience-building strategy. In response, and as part of its governmental mandate, the GoL initiated public–private partnerships (mainly with the Bank) for climate financing with the primary aim of enhancing these citrus farmers' adaptation strategies. Despite this noble GoL aim in the context of the climate emergency, the implementation modalities of the partnership have been fraught with complex ideological, financial, operational and logistical implementation challenges as well as omissions that have derailed its vision of creating a pathway towards climate justice.

Firstly, the public–private partnership model is ideologically informed by discredited neoliberal thinking and actions. By enticing smallholder farmers to increase the quantity and quality of their citrus fruits through cultivation of every inch of available land, the Bank and GoL are directly promoting unsustainable land use in pursuit of profit. In the long term, land quality, soil nutrients and productive capacity will decline, thus defeating the primary goal of building farmer resilience as a pathway towards social and climate justice. Secondly, climate finance raised and distributed to farmers through inputs, technical and technological support is inadequate to address underlying drivers of vulnerability and production challenges. Farmers thus continue to bear the brunt of climate impacts, suggesting climate finance has not set them on a sustainable path towards distributive justice. Thirdly, even though the GoL and the Bank occasionally observe procedural justice through involving farmers in the planning and implementation of funded climate actions, this is somewhat overshadowed by operational and logistical challenges of the public–private partnership. For example, bureaucratic 'red tape' in government departments, frequent government paperwork completion and submission by farmers to enable access to financial and technical resources and prescriptive cropping choices and patterns by the Bank all undermine elements of procedural justice in the public–private partnership. Lastly, despite these limitations of the public–private partnership, it has provided a starting point that can be transformed for effectiveness and sustainability into the future. This transformation should involve raising more grant climate finance from other national and international sources to curtail reliance on the neoliberal bank. Once raised, it should be equitably and fairly disbursed to address poverty and vulnerability drivers bedevilling the smallholder farming sector as an avenue towards distributive justice. As a pathway towards climate justice, active participation and farmer representation in decision-making platforms (at the policy and programming level) is crucial. Such procedural justice is important because it will enhance usually underrepresented vulnerable smallholder farmers, with the least capacity to build resilience, in planning and implementation of climate resilience actions towards climate justice.

References

Adams, B. and Luchsinger, G. (2009) 'Climate Justice for a Changing Planet: A Primer for Policy Makers and NGOs' (available at: https://unctad.org/system/files/official-document/ngls20092_en.pdf).

Aglietta, M., Hourcade, J., Jaeger, C. and Fabert, B.P. (2015) 'Financing Transition in an Adverse Context: Climate Finance Beyond Carbon Finance', *International Environmental Agreements: Politics, Law and Economics*, 15, 403–420.

Barrett, S. (2014) 'Subnational Climate Justice? Adaptation Finance Distributions and Climate Vulnerability', *World Development*, 58, 130–142.

Boissinot, J., Huber, D. and Lame, G. (2016) 'Finance and Climate: The Transition to a Low-Carbon and Climate-Resilient Economy from A Financial Sector Perspective', *Financial Market Trends*, 1, 7–23.

Bowen, K., Cradock-Henry, N. and Koch, F. (2017) 'Implementing the Sustainable Development Goals towards Addressing Three Key Governance Challenges: Collective Action, Trade-offs and Accountability', *Current Opinion in Environmental Sustainability*, 26, 90–96.

Bracking, S. (2015) 'The Anti-Politics of Climate Finance: The Creation and Performativity of the Green Climate Fund', *Antipode*, 47(2), 281–302.

Brown, J., Bird, N. and Schalatek, L. (2010) 'Climate Finance Additionality: Emerging Definitions and Their Implications', *Climate Finance Policy Brief*, No. 2 (available at http://citeseerx.ist.psu.edu/viewdoc/download?doi=10.1.1.626.7000&rep=rep1&type=pdf).

Chinoswky, P., Hayles, C., Schweikert, A., Strzepek, N., Strzepek, K. and Schlosser, C.A. (2010) 'Climate Change: Comparative Impact on Developing and Developed Countries', *Engineering Project Organization Journal*, 1(1), 67–80.

Colenbrander, S., Dodman, D. and Mitlin, D. (2018) 'Using Climate Finance to Advance Climate Justice: The Politics and Practice of Channelling Resources to the Local Level', *Climate Policy*, 18(7), 902–915.

Corsatea, T.D., Giaccaria, S. and Arantegui, R.L. (2014) 'The Role of Sources of Finance on the Development of Wind Technology', *Renewable Energy*, 66, 140–149.

Fisher, S. (2015) 'The Emerging Geographies of Climate Justice', *The Geographical Journal*, 181(1), 73–82.

Gampfer, R., Bernauer, T. and Kachi, K. (2014) 'Obtaining Public Support for North-South Climate Funding: Evidence from Conjoint Experiments in Donor Countries', *Global Environmental Change*, 29, 118–126.

Grasso, M. (2009) 'An Ethical Approach to Climate Adaptation Finance', *Global Environmental Change*, 20, 74–81.

Gwimbi, P. (2017) 'Mainstreaming of Adaptation Programmes of Action into National Development Plans in Lesotho: Lessons and Needs', *International Journal of Climate Change Strategies and Management*, 9(3), 299–315.

Hallegatte, S., Bangalore, M., Bonzanigo, L., Fay, M., Kane, T., Narloch, U., Rozenberg, J., Treguer, D. and Vogt-Schilb, A. (2016) *Shock Waves: Managing the Impacts of Climate Change on Poverty*. Washington, DC: World Bank (available at http://hdl.handle.net/10986/22787).

Harvey, B., Jones, L., Cochrane, L. and Singh, R. (2019) 'The Evolving Landscape of Climate Services in Sub-Saharan Africa: What Roles Have NGOs Played?', Climate *Change*, 157, 81–98.

Huq, S., Reid, H., Konate, M., Rahman, A., Sokona, Y. and Crick, F. (2011) 'Mainstreaming Adaptation to Climate Change in Least Developed Countries', *Climate Policy*, 4(1), 25–43.

Huang J. and Wang, Y. (2014) 'Financing Sustainable Agriculture under Climate Change', *Journal of Integrative Agriculture*, 13(14), 698–712.

Kameyama, Y., Morita, K. and Kubota, I. (2016) 'Finance for Achieving Low-Carbon Development in Asia: The Past, Present and Prospects for the Future', *Journal of Cleaner Production,* 128(2016), 201–208.

Khan, M., Robinson, S., Weikmans, R., Ciplet, D. and Roberts, J.T. (2020) 'Twenty Five Years of Adaptation Finance Through a Climate Justice Lens', *Climatic Change*, 161, 251–269.

Lesotho Meteorological Services (2013) *Lesotho Climate Change Baseline and Trend Analysis Report: Towards a Resilient Future*. Maseru: Lesotho Meteorological Services.

Mahat, L., Blaha, L., Uprety, B. and Bitter, M. (2019) 'Climate Finance and Green Growth: Reconsidering Climate-Related Institutions, Investments and Priorities in Nepal', *Environmental Sciences Europe*, 31(46), 24–36.

Ministry of Energy, Meteorology and Water Affairs (2013) *Lesotho Vulnerability Assessment and Analysis Report*. Maseru: Ministry of Energy, Meteorology and Water Affairs.

Ministry of Trade, Cooperatives, Industry and Marketing (2017) *Lesotho Diversification Support Project: Appraisal Report*. Maseru: Ministry of Trade Cooperatives Industry and Marketing.

Pauw, P. and Bendandi, B. (2016) 'Remittances for Adaptation: An Alternative Source of International Climate Finance. Migration, Risk Management and Climate Change', *Evidence and Policy Responses*, 6, 195–211.

Perkins, R. and Nachmany, M. (2019) 'A Very Human Business: Transnational Networking Initiatives and Domestic Climate Action', *Global Environmental Change*, 54, 250–259.

Porter, L., Richards, L., Verlie, B., Bosomworth, R., Moloney, S. and Ben, B. (2020) 'Climate Justice in a Climate Changed World', *Planning Theory and Practice*, 21(2), 293–321.

Rojas-Downing, M.M., Nejadhashemi, A.P., Harrigan, T. and Woznicki, S.A. (2017) 'Climate Change and Livestock: Impacts, Adaptation and Mitigation', *Climate Risk Management*, 16, 145–163.

Schlosberg, D. and Collins, L.B. (2014) 'From Environmental to Climate Justice: Climate Change and the Discourse of Environmental Justice', *WIREs Climate Change*, 5(5), 359–374.

Sithole, A. and Murewi, C.T.F. (2009) 'Climate Variability and Change over Southern Africa: Impacts and Challenges', *African Journal of Ecology*, 47(1), 17–20.

Stalley, P. (2018) 'Norms from the Periphery: Tracing the Rise of the Common but Differentiated Principle in International Environmental Politics', *Cambridge Review of International Affairs*, 30(2), 141–161.

Steckel, J.S., Jakob, M., Flachsland-Kornek, U., Lessmann, K. and Edenhofer, O. (2016) 'From Climate Finance toward Sustainable Development Finance', *WIREs Climate Change*, 8(1), 34–45.

The Nature Conservancy Climate Change Program (2012) 'Climate Finance Readiness: Lessons Learned in Developing Countries Report', (available at https://www.nature.org/media/climatechange/climate-finance-readiness.pdf)

Urpelainen, J. (2012) 'Strategic Problems in North–South Climate Finance: Creating Joint Gains for Donors and Recipients', *Environmental Science and Policy*, 21, 14–23.

Vanderheiden, S. (2015) 'Justice and Climate Finance: Differentiating Responsibility in the Green Climate Fund', *The International Spectator*, 50(1), 31–45.

Warren, P. (2019) 'The Role of Climate Finance Beyond Renewables: Demand-Side Management and Carbon Capture, Usage and Storage', *Climate Policy*, 19(7), 861–877.

Yamineva, Y. (2016) 'Climate Finance in the Paris Outcome: Why Do Today What You Can Put Off Till Tomorrow?', *Review of European, Comparative and International Environmental Law*, 25(2), 174–185.

6 Just Transition and Sustainability

Implications for Poverty, Inequality and Jobs in Eswatini

Sipho F. Mamba and Thabo Ndlovu

Introduction

The sustainability transition discourse has, in most recent years, attracted the attention of policymakers and massive research due to its perceived usefulness in promoting socio-technical transformation towards a more sustainable future (Ramos-Mejía et al., 2018). The ultimate desire by member states (and countries in general) to fulfil the dictates of the United Nations Agenda 2030 on Sustainable Development has not been without influence. However, the sustainability transition call has been received with mixed feelings in the Global South and in Eswatini, in particular (Martiskainen & Sovacool, 2021; Yazdani & Dola, 2013), as the discourse has not been without a political connotation to the country (Avelino et al., 2016; Geels, 2014) as it continues to spark major debates between the North and South countries due to the different lenses through which this transition call is viewed.

The sustainability transition literature in the Global South, Eswatini included, has, to a larger extent, focused on providing illumination on the co-evolution of socio-technical transitions (Hielscher et al., 2022; Loorbach et al., 2017), with a strong focus on sustainable ways of production and consumption (Köhler et al., 2019; Magnusson & Werner, 2022; Markard et al., 2012; Smith et al., 2010). Scholars have been more concerned about the nexus between technological innovation diffusion and societal systems, especially how the systems (e.g. energy systems) change with technological innovations and, most recently, the cessation of unsustainable innovations and incumbent elements, destabilization of dominant pathways for a sustainable society (Figenbaum, 2017; Johnstone & Stirling, 2020).

Adaptation efforts remain poorly funded relative to mitigation projects, both globally and in Africa specifically. Less attention has been given to the 'socio' part of the socio-technical and/or human dimension of the transition, more so in Eswatini (Martiskainen & Sovacool, 2021; Zolfagharian et al., 2019). As such, the sustainability transition and poverty cocktail, coupled with the need for job creation and reduction of inequality gap in Eswatini, are predominantly under-researched areas within the sustainability transition discourse. Yet, issues of job creation and narrowing of the gap between the haves and have-nots remain pertinent and should form part of the drive towards a sustainable society, if the transition

DOI: 10.4324/9781003397120-8

is to be 'just'. Regarding the just transition processes in the Global South nations such as Eswatini are the poorly funded adaptation efforts compared to mitigation projects (Ziervogel *et al.*, 2022). It is in the light of this omission that Ramos-Mejía *et al.* (2018) call for an integration of the poverty alleviation initiatives into the sustainability transitions agenda in countries of the Global South. Through a systematic review of scholarly literature, this chapter engages with the sustainability transition discourse and draws attention to the contentious geopolitics of sustainability transitions in the context of compounding factors such as poverty, inequality and unemployment in the Kingdom of Eswatini.

Sustainability and the Just Transition: A Reflection on Frameworks

The 'sustainability transition' concept can be better understood when traced back from the idea of 'sustainable development' which the World Commission On Environment and Development (1987) views as 'development that meets the needs of the present generation without compromising the ability of future generations to meet their own needs'. Since the propagation of these iconic documents: 'Our Common Future' and 'Limits to Growth', the concept of 'sustainability' entered the lexicon and development agendas of many countries and has sparked 'hot' debates, globally. The Earth Summit of 1992, held in Rio de Janeiro, has not been without influence in the current global debates on this concept, and it marked the initiation of the global agenda (Agenda 21) which became a charter for achieving sustainable development (Echendu & Georgeou, 2021).

Just like any other concept, 'sustainability' has attracted different scholarly debates and academic positions on its definition. However, an existing general consensus of this concept is that for any development to be considered viable, it should use a future lens as it strives to balance current economic needs and environmental protection. In Allen's view (1980), for instance, any development to be considered sustainable, it should satisfy the needs of the present generation and improve its quality of life in the confines of the available natural resources, without exerting too much strain on it and the utilized ecosystem, for both to be able to regenerate. In other words, sustainable development is, actually, striking a balance between environmental protection, social development and economic growth: 'people, planet and prosperity', commonly referred to as 'the triple bottom line' (Hammond, 2000; Parkin, 2000) – which provides a connection between 'quality of life' and 'ecosystem conservation' (Baumgärtner & Quaas, 2010; Yazdani & Dola, 2013).

The debate on environmental protection is incomplete without interrogating climate change and the just transition nexus. Climate change has become one of the major challenges we face (Satgar, 2018), and the proposed interventions are under scrutiny in the transition lens for embracing views of those bearing the brunt of climate change impacts. This is despite the fact that other scholars such as Harding *et al.* (2022) posit that transition has become a buzzword among policymakers and practitioners. Such a notion questions the significance of the just transition towards shifting from dirty energy to energy democracy and from rampant destructive development to ecosystem restoration (Alliance,

2018). In contemporary development, this concept is inescapable in describing where we are going and how we get to effectively address the climate action injustices as well as ensuring that no one is left behind. The continued rise in global anthropogenic greenhouse gas (GHG) emissions, which causes climate change, accentuates the importance of just transition in shaping energy policies towards balancing socioeconomic and ecological considerations in response to climate crisis (Bainton *et al.*, 2021). This confirms the long-standing view by Elkington (1997) that for any development to be sustainable, it should cover 'the triple bottom line' by integrating social justice and environmental concerns to any form of economic development. As such, the Kingdom of Eswatini, through the Eswatini Environmental Authority, has made it mandatory for development projects to be subjected to environmental and social impact assessment. Assessments, as argued by Jenkins *et al.* (2020), help unpack the burdens of climate action and the social distribution cost and benefits linked to energy transition. The popularity of the just transition heightens growing awareness about deepening inequalities between the rich and poor of the world (Bainton & McDougall, 2021); hence, inequalities generated by efforts to address climate crisis cannot be ignored (Svobodova *et al.*, 2020).

Against this background, it should suffice to define the 'sustainability transition' concept as "fundamental changes in socio-technical systems such as energy, food or transport that aim to address grand challenges in a way that meets the needs of the present without compromising the ability of future generations to meet their own needs" (Markard *et al.*, 2020, p. 081001). It entails a shift, usually from a polluting socio-technical system to a more sustainable one (i.e. shift from fossil fuel [nonrenewable] to solar [renewable] energy), which may necessitate a change in regulations, practices and institutions (Martiskainen & Sovacool, 2021; Schot & Geels, 2008). This suggests that sustainability transition is nothing else, but the transformation of socio-technical systems towards a sustainable society in response to multiple and persistent problems that find expression in contemporary societies (Grin *et al.*, 2010). 'Transition' is, therefore, perceived by the proponents of the sustainable transition discourse as either a process (towards achieving a sustainable society) or desired outcome of sustainability (Silva & Stocker, 2018).

Scholars confirm the inextricable link between poverty, inequality, economic growth and sustainable development (Romijn *et al.*, 2010; Zhu *et al.*, 2022). In Zhu *et al.*'s words (2022, p. 27613), "inclusive growth, poverty alleviation, and sustainable development are associated with each other, while poverty reduction is considered a sustainable development principle". Unfortunately, the link between sustainable development theory and poverty alleviation remains largely unclear (Leal Filho *et al.*, 2022). This is partly due to poor integration of poverty alleviation efforts to sustainable development practices. Nonetheless, the constellation of poverty, inequality and economic growth demand attention if the target to meet the sustainable development goals (SDGs) agenda is to be realized by 2030, particularly in the Global South (United Nations, 2022). Romijn *et al.* (2010), however, acknowledge the challenge of connecting environmental sustainability with poverty reduction agendas, declaring this as a 'missing link' in sustainability transition studies, which has resulted in limited understanding of this critical connection to holistically understand the sustainability transition dynamics.

It is through this connection that the idea of the 'just transition' notion was born in the early 1970s. Sustainability transition history has it that the 'just transition' concept was given birth to by the trade unionist, Tony Mazzocchi, in his determination to reconcile social and environmental concerns (Stevis *et al.*, 2022). In other words, for a transition to be 'just' and 'sustainable', it should consist of a dual commitment to sustainability and human well-being (Sabato & Fronteddu, 2020; Swilling *et al.*, 2016). This introduces the concept of 'social justice' in transition which tends to characterize current sustainability literature (Bastos Lima, 2022; Benitez, 2018; Stevis *et al.*, 2022). Proponents of social justice in the transition space advocate for attention to be paid to social justice concerns to ensure that environmental protection and economic growth do not occur at the expense of people or social equity (Benitez, 2018; Cohen, 2020). It is, however, important to understand and appreciate the theoretical frameworks that guide and explain transitions in the sustainability space.

Several theoretical frameworks are used in the field of sustainability transition, and these include the transition management (TM), strategic niche management (SNM), the technological innovation system (TIS) approach and the multi-level perspective (MLP) (Kinn, 2016; Markard *et al.*, 2012). The TIS approach, to begin with, is a prominent framework in the sustainability transition space which borrows largely from the industrial and innovation system theories (Markard *et al.*, 2015; Weber & Truffer, 2017). This approach pays less attention on stabilizing existing systems but rather focuses, to a larger extent, on the emergency of novel innovations and new technological development (Bergek *et al.*, 2008; Köhler *et al.*, 2019).

The SNM is a recently crafted method predominantly used to analyse and facilitate diffusion of fundamentally new sustainable innovations through experiments. The agriculture commercialization project in the Lowveld of Eswatini provides a good example of the SNM approach where sugarcane demonstration plots have been set up in order to improve livelihoods and socioeconomic status of households in drought-prone areas of the country. This approach holds that radical innovations are developed in 'protected spaces', and niche innovations are usually set up through interactions between social networks, learning processes and expectations (Kemp *et al.*, 1998). According to the SNM approach, sequences of experiments and demonstration projects enable recursive cycles of these processes, which can generate innovation trajectories (Geels & Raven, 2006). In this framework, the character and shape of innovation trajectories is influenced largely by social networks, expectation and learning (Schot & Geels, 2008). The TM is a multi-level model of governance and a policy-oriented framework which shapes processes of co-evolution (and is also referred to as 'co-evolutionary management') through the use of transition experiments, visions and cycles of adaptation and learning (Köhler *et al.*, 2019). The framework allows policymakers to shape transitions through some sequential steps: strategic activities (i.e. determining possible transition pathways), tactical activities (i.e. plan development), operational activities (i.e. experiments) and reflexive activities (i.e. project monitoring and evaluation) (Loorbach & Rotmans, 2010).

The MLP adopts an integrative approach to understanding transitions as resulting from co-evolutionary processes and has been useful in the analysis of socio-technical transitions in the sustainability space. The MLP framework considers transitions to result from the interaction of three processes: niches, socio-technical regimes and socio-technical landscape (El Bilali, 2019; Smith *et al.*, 2010). Niches are believed to be a source of radical innovation and provide an 'incubation' for market forces, allowing for research and learning through experience. The breakthrough of any niche innovation depends largely on the pressure put by landscape development on the regime that leads to tension, cracks and window of opportunity. The interaction between regimes and niches occurs on multiple dimensions which may include technology dimension, culture and regulation dimension (Köhler *et al.*, 2019). The MLP is regarded as the most prominent and widely used approach in transition studies (El Bilali, 2019).

Although the MLP and SNM are the most prominent and widely used sustainability approaches, Ramos-Mejía *et al.* (2018) argue that these approaches have a strong northern influence and, hence, are actually less relevant for use in the Global South. They further note that the limitations of using these approaches in the South have not been explored in any detail, probably due to the sustainability literature that tend to be biased towards the North, raising an urgent need to engage with the Global South (Markard *et al.*, 2020). Needless to say, scholars have observed that the just transition literature reflects a strong bias towards environmental considerations in the North (which may be less of a priority in the South) and so are the socio-technical transformations which are, more often than not, aimed at climate change alternatives in the European countries (Climate Justice Alliance, 2022).

While environmental protection is also important in the South, welfare issues are even more important, creating distinct 'sustainability focus' between the North (which focus mainly on environmental protection) and South (which focus mainly on social issues such as poverty reduction and employment creation, etc.). As Blowers and Pain (1999) rightly note, what is considered a need in the South may be luxury in the North, and therefore, determining the same sustainability goals and applying the same assessment criteria for progress in both the South and North would certainly be debatable (Yazdani & Dola, 2013), especially because sustainable development is time and spatial dependent, and thus, standardized universal objectives may not achieve much compared to context-specific strategies.

Just Transition and Sustainable Development Goals

Scholars acknowledge the existence of a connection between the just transition and the SDGs (Climate Justice Alliance, 2022; European Bank, 2022; Pollin, 2021; Yazdani & Dola, 2013). In fact, the European Bank (2022) documents that the just transition concept links perfectly well to the SDGs, explicitly drawing together SDGs 7 (affordable and clean energy), 8 (decent work and economic growth),

10 (reduced inequalities) and 12 (climate action), among others. The analysis of the Paris Agreement reflects the connection by highlighting the importance of the 'just transition' on the provision of quality jobs which, according to the agreement, should align with national development priorities.

We can claim that the SDGs provide a unique and powerful unified programme, not only for environmental protection but also for improving the living standards through poverty reduction and employment creation (Pollin, 2021). The global migration from coal-powered energy systems to green energy sources in an effort to reduce carbon footprint is one clear sign of efforts to ensure a more sustainable society and way of promoting socio-technical transformation towards a more sustainable future (Ramos-Mejía *et al.*, 2018). As Pollin (2021) rightly observes, the reduction of carbon dioxide (CO_2) emissions (in an effort to address climate change issues) is a major focus of the sustainability call, especially because about 70% of the global energy is derived from this less desirable fossil fuel in the sustainability space.

While the transition and transformation of socio-technical systems may be viewed as positive in some cases (resulting in positive results), scholars also observe that such changes may also produce negative effects, particularly in the Global South countries such as the Kingdom of Eswatini. The different levels of growth and development (which also determine the priority areas) between the Global North and South may account for the difference in the impacts of the transition on these countries. The advent and interruption by the COVID-19 global pandemic, coupled with the current worldwide economic recession (intensified by the Russian–Ukrainian war), slowed down the pace of poverty alleviation, and this has an implication on the sustainability transition commitments and targets globally and in developing countries, in particular where poverty is claimed to have deep-rooted in 2020 (Zhu *et al.*, 2022).

Just Transition Context in Eswatini

The Kingdom of Eswatini is one of the countries in sub-Saharan Africa which embraces the United Nations Agenda and continues to confirm its commitment towards the achievement of the global 2030 Agenda and to its inclusive partnership approach aimed at achieving the SDGs. As a means to embrace and domesticate the UN 2030 Agenda, the Kingdom of Eswatini integrated the SDGs to its National Development Strategy (NDS) and Vision 2022. These important national documents present an opportunity for the integration of the 2030 Agenda and other emerging and pressing development issues of national importance, simultaneously fostering a perfect alignment between the NDS and SDG targets (Eswatini Government, 2019).

As an active partner to the commitment of achieving the SDGs by 2030, the Kingdom of Eswatini has domesticated the SDGs and identified ten national SDG priorities aimed to guide its development paths (United Nations – Eswatini, 2021). To date, the country has made progress on SDGs 3, 7, 13 and 17 with less achieved on the other prioritized goals. It is important to note that the Kingdom of Eswatini is currently prioritizing SDG 1 where the country is committed to end poverty in all its forms through creating employment opportunities to reduce social and economic inequalities (United Nations – Eswatini, 2021). The country rolled out several programmes which include the agriculture commercialization project (which encourages rural farmers to form farmer associations and grow sugarcane for the country's booming sugar industry) and capital provision (through the Regional Development Fund [RDF]) to rural associations for several development projects.

The country, through its ambitious policies, is committed to transforming the energy sector and system to reduce its reliance on neighbouring South Africa, simultaneously taking into account issues of poverty and inequality that find expression in the country (Roth *et al.*, 2022). While the country is committed to following sustainable and recommended green energy paths, its economic landscape forces it to prioritize thermal power (thermal power station under construction) whereas attending to other pressing national issues such as poverty alleviation and employment creation. This scenario constrains national government resources and that of other partners, and it impedes their ability to balance their acts in terms of addressing socioeconomic issues and environmental protection issues.

The middle-income country status (which creates an inability for the country to access concessional loans) pauses as a great challenge for the country towards implementation of the SDGs, forcing the country to prioritize economic growth and strengthening of social sectors over the promotion of environmental sustainability (Eswatini Government, 2019). The onset of the COVID-19 global pandemic in the country (March, 2020) further aggravated pre-existing socioeconomic inequalities, pausing a serious threat to economic progress and realization of the 2030 Agenda (United Nations – Eswatini, 2021). It is on the basis of these constraining factors that the Kingdom of Eswatini's development agenda is more inclined towards economic recovery, although also recognizing the importance of investing in environmental sustainability. The country's national strategy for sustainable development and inclusive growth and the strategic road map for the year 2019–2022 (which guide development and economic recovery for the Kingdom of Eswatini) also prioritize human development, placing more emphasis on employment creation, poverty alleviation and gender equity and social integration (WFP, 2019).

Furthermore, the Kingdom of Eswatini prioritizes environmental protection and, as such, developed a Strategy for Sustainable Development and Inclusive Growth (2017) which mainstreams the 2025 (Southern African Development Community (SADC) Agenda, African Union Agenda 2063 and the UN Sustainable Development Agenda 2030 (Eswatini Ministry of Economic Planning and Development, 2017). However, poverty alleviation remains the key priority area of focus, owing to its high unemployment and poverty rates (with national poverty rate at 58.9% and extreme poverty at 20.1%), thus the NDS was operationalized by the Poverty Reduction Strategy and Action Plan (PRSAP) of 2008 (World Bank, 2022). The country's strategic road map (2019–2023) medium-term development framework also prioritizes key sectors with the aim to expedite economic growth to ensure financial stability and improvement in the quality of life.

Just Transition Implications on Employment and Poverty Reduction in Eswatini

The changes in socio-technical systems associated with the 'just transition' concept aimed at producing a better world, undoubtedly have implications on the SDGs of the United Nations (Pollin, 2021; Yazdani & Dola, 2013). The global migration from coal-powered energy systems (on which Eswatini relies) to green energy sources in an effort to reduce carbon footprint, although a good initiative in environmental terms, has an implication on employment opportunities globally as more than 70% of the global energy is derived from fossil fuel which is a less desirable energy source in the sustainability space (Ritchie *et al.*, 2022). Developing countries such as the Kingdom of Eswatini feel the hardest blow of this transformation as the world promotes socio-technical transformation towards a more sustainable future in the energy sector (Ramos-Mejía *et al.*, 2018). The impact of this transition on jobs and employment opportunities in Eswatini which still grapples with flattening the high unemployment rate curve remains a major concern. Unfortunately, the reduction of CO_2 emissions (in an effort to address climate change issues and reach climate neutrality) is the major focus of the sustainability narrative/call (Pollin, 2021) which implies the inevitable disruption of the energy sector and the implication on job opportunities. Needless to say, the country's effort to reduce its reliance on South Africa for electrical energy (where it imports over 70%) through the planned establishment of a thermal power plant (with employment prospects) seems to be less supported by the energy transition call which emphasizes on 'clean' energy which is less affordable for low economies such as the Kingdom of Eswatini. It is argued that without transforming the energy sector globally, it will not be possible to achieve the 1.5-degree target of the Paris Agreement (IPPC, 2022).

Developing countries already grappling with high unemployment rates such as the Kingdom of Eswatini, the phasing out of thermal power plants for green energy (e.g. solar power) would undoubtedly have detrimental consequences. The transition (which is supposed to be just) will perhaps increase informal employment in Eswatini (which currently stands at 65.2%) and subsequently increase the already high rural poverty in the country (which currently stands at 70.2%) where the majority rely on remittances from such industries (Eswatini Government, 2019). This retards progress made towards meeting the Agenda 2063 targets, especially Goal 4 on Transformed Economies and Job Creation in order to achieve annual Gross Domestic Product (GDP) growth (of at least 7%) (Eswatini Government, 2019).

It is argued that although investment in the transition to green energy will, in the long run, generate sustained jobs for millions of people (two to three times more jobs than continued investment in the current fossil fuel energy system), millions of workers will lose jobs in the fossil fuel phase out. In addition, millions of people whose livelihoods depend on coal, oil and natural gas will also be affected. Unless the transition is just, it might "act counter to the SDGs, creating winners and losers, and heightening inequality" (Pollin, 2021, p. 1). If the process of transition is not just, the outcome will never be (European Bank, 2022).

Poverty and Inequality

'Leave no one behind' is the guiding principle underlying the implementation of Agenda 2030 of the United Nations (and its SDGs), which also applies to the 2063 African Agenda, both of which emphasize on poverty eradication and reduction of inequality. This key development objective is also highlighted in the country's vision (Vision 2022) just like it is in other country's development objectives (i.e. Botswana's Vision 2036, Zimbabwe's Vision 2030, Lesotho's Vision 2030) (Government of Botswana, 2018; Government of Lesotho, 2019; Government of Zimbabwe, 2018).

In the already precarious economic landscape of Eswatini and of sub-Saharan Africa in general, further reduction in employment opportunities heightens the rate of poverty and exacerbates economic and social inequalities. Over two million of the population in Africa is employed in thermal power industries (International Renewable Energy Agency [ERENA], 2022), and in the transition period, this proportion will shrink and contribute to limited employment opportunities which further deepen poverty and inequality in Africa. Due to the low local employment opportunities in Eswatini, most Swazis work in thermal power production industries in neighbouring countries (e.g. South Africa) and support their rural relatives through remittance flow. Drying up this important income channel (which is very important in poor rural spaces) will inevitably heighten rural poverty and increase the already high inequality gap (with Gini coefficient of 0.51) (United Nations – Eswatini, 2023).

Poverty reduction is a critical issue and great development challenge confronting the Kingdom of Eswatini and sub-Saharan Africa, in general. The prevailing social and economic inequalities born out of unequal distribution and allocation of resources in Eswatini erode the ability of its citizens to meet the needs and contribute to the just transition agenda. However, as already noted, the country had paid specific focus to levelling the poverty curve; hence, poverty alleviation is one of the country's development goals which corresponds with Goal 1 of Agenda 2063 of achieving a high standard of living, improved quality of life and well-being for all.

The Kingdom of Eswatini recently launched its economic blueprints to guide the 2019–2021 Strategic Road Map and the 2019–2023 National Development Plan (NDP) (United Nations – Eswatini, 2021). Both of these important national documents underscore the importance of economic recovery and the improvement of the quality of life of the Swazi nation, especially after the onset of the devastating global pandemic, COVID-19, which has undermined the country's efforts and progress in reducing the proportion of the poor (United Nations – Eswatini, 2021). Among the key priorities of the country as enshrined in the country's PRSAP is the desire to eradicate poverty by 2022 which necessitated a change in the country's development priorities and approaches (Eswatini Government, 2007).

Policy and Just Transition in Eswatini

Mcneill (2005) contends that there is relatively lack of attention given to economic policy initiatives aimed at smoothing the progress towards a sustainable society in Africa (just transition). Where minor attempts have been made to discuss economic policy issues in the context of the just transition, such policies tend to be very abstract without proper implementation considerations. If the transition is to be 'just' for African societies, sustainability policies and practices should incorporate key issues of national importance such as poverty eradication and creation of employment opportunities. In fact, scholars argued that "sustainability sits at the nexus of poverty, the natural environment and innovation" (Khavul & Bruton, 2013, p. 287) and that "a just transition should consist of a dual commitment to human well-being" (Swilling *et al*., 2016, p. 650). This highlights the centrality of socioeconomic issues in the transition debate, if the transition is to be 'just'.

In addition to the integration of pertinent issues of national importance in the just transition framework at local level, it is pertinent that an enabling policy environment is created for the full implementation of the 'just transition'. The Kingdom of Eswatini has provided an enabling policy environment for adherence to the Sustainable Development Conventions (e.g. the Paris Agreement and Sendai Framework for Disaster Risk Reduction) (Eswatini Government, 2019). The country has localized the SDGs by crafting several policy documents for such purpose. These include the NDP (2019–2022), the Strategic Roadmap (2018–2023), Strategy for Sustainable Development and Incisive Growth (SSDIG) – NDS review (2014) and the NDS (Vision 2022) (Eswatini Government, 2020). These documents provide guidelines and opportunities for sustainable economic growth and environmental protection. Among the key focus areas in the country's development plan is to manage natural resources and ensure environmental sustainability, enhance social and human development, ensure efficient public service delivery and support sustainable inclusive growth, among others. This highlights the integration of sustainable development issues in the country's development plan (Eswatini Government, 2020) and the creation of an enabling environment for the just transition.

The country has also operationalized the 'Leaving No One Behind' principle by setting up a social protection intervention policy. Through this policy, several social protection initiatives have been set up, and these include the elderly and disabled grant, Orphans and Vulnerable Children grant (OVCs), the RDF for Rural Development and the Youth Enterprise Fund and Medical Fund (Phalala) (Eswatini Government, 2020). While the elderly/disabled and OVC grants provide financial support to individuals to enable them to meet their basic and educational needs, the RDF and Youth Fund provide capital for Small and Medium Enterprises (SMEs) in rural areas. The Phalala medical grant helps in paying medical bills. While providing an enabling policy environment is critical for the achievement of just transition targets, there is need for proper integration of socioeconomic and environmental sustainability issues to key development programmes for a just and effective transition.

Conclusion

The different levels of growth and development (which also determine the priority areas) between the Global North and Global South countries such as the Kingdom of Eswatini largely account for the differences in the impacts of the transition on these countries. While environmental issues are key focal areas of sustainability in the North, social issues such as poverty alleviation and employment are a major concern in developing states, especially in Eswatini and, therefore, remain the key focus areas in the country, just like in other Global South countries (the green versus brown agenda) (Ayoo, 2022). The advent of the COVID-19 global pandemic (and its associated interruptions), coupled with the current worldwide economic recession (fuelled by the Ukraine invasion), slowed down the pace of poverty alleviation, and this has an implication on the sustainability transition commitments and targets globally and in Eswatini, in particular, where poverty is deep-rooted (Zhu *et al.*, 2022). The Kingdom of Eswatini is among the sub-Saharan countries with high poverty rates, coupled with high rates of unemployment and high levels of inequalities. These constellations demanded urgent policy actions and special attention; hence, these are taking priority in the country's development agenda (UN-Eswatini, 2021). The country's development path and associated policy documents (e.g. NDS, national strategy for sustainable development, PRSAP and the country's Vision 2022) attest to the prioritization of poverty alleviation, job creation and reduction of inequalities (SDGs 1, 8 and 10) (Eswatini Government, 2019; UNDP, 2023; WFP, 2019). This pattern seems to characterize most countries with a similar development context where social and economic developmental issues are a pressing need.

Poverty eradication and combating inequality are part of the sustainable development agenda and are, in actual fact, fundamental for a just and safe space for humanity (Connors *et al.*, 2018) and, therefore, an important aspect of the just transition call. The just transition initiatives for developing countries should be characterized by the integration of key socioeconomic issues aimed at combating poverty and unemployment which are key developmental issues in the South. The focus on climate change, with a parallel neglect of other pressing issues in developing countries, such as poverty and unemployment, can reverse or erase many development achievements. This includes improvements in living conditions which may have been achieved in several decades of development efforts and initiatives (Connors *et al.*, 2018). In this connection, some countries in the South such as the Kingdom of Eswatini have intentionally prioritized socioeconomic issues (e.g. poverty eradication and employment creation) in the midst of compounding factors such as COVID-19 and economic meltdown which, collectively, account for the prevailing economic landscape.

While the sustainability transition is a plausible global initiative meant to yield positive results with regard to the provision of green jobs, the phasing out of carbon-intensive activities has the propensity to result in numerous job losses, and its effects felt more in developing nations (Sharpe & Martinez-fernandez, 2021). These are some of the social realities associated with the just transition which should not be ignored while we get absorbed in the task of transformation. As Mcneill (2005) has argued, giving limited attention to the sensitivity of the current transition with regard to job losses is most likely to jeopardize the attainment of a sustainable society, particularly in developing countries which are characterized by high rates of poverty and unemployment which continue to pause as major developmental challenges. Sustainability issues, although important, may not be a priority in the Global South countries, not only due to their debt burden and fiscal situation but also because of some pressing socioeconomic issues that have forced themselves into the development agenda of these countries (Pickett *et al.*, 2013; Thring 1990). The creation of an enabling policy environment for proper integration of socioeconomic issues and environmental protection in key development programmes remains crucial for a smooth and just transition.

References

Allen, R. (1980). *How to Save the World.* Barnes and Noble.
Alliance, C. J. (2018). Just Transition Principles. Accessed 13/02/2023 from https://climate-justicealliance.org.
Avelino, F., Grin, J., Pel, B., & Jhagroe, S. (2016). The Politics of Sustainability Transitions. *Journal of Environmental Policy and Planning, 18*(5), pp. 557–567. https://doi.org/10.10 80/1523908X.2016.1216782.
Ayoo, C. (2022). Poverty Reduction Strategies in Developing Countries. In P. de Salvo, & M. V. Piñeir (Eds.), *Rural Development-Education, Sustainability, Multifunctionality* (pp. 17–57). London: IntechOpen.
Bainton, N., Kemp, D., Lèbre, E., Owen, J. R., & Marston, G. (2021). The Energy-Extractives Nexus and the Just Transition. *Sustainable Development, 29*(4), pp. 624–634.
Bainton, N., & McDougall, D. (2021). Unequal Lives in the Western Pacific. In N.A. Bainton, D. McDougall, K. Alexeyeff, & J. Cox (Eds.), *Unequal Lives: Gender, Race and Class in the Western Pacific* (pp. 1–46). Canberra: ANU Press.
Bastos Lima, M. G. (2022). Just Transition towards a Bioeconomy: Four Dimensions in Brazil, India and Indonesia. *Forest Policy and Economics, 136*(July 2021), p. 102684. https://doi.org/10.1016/j.forpol.2021.102684.
Baumgärtner, S., & Quaas, M. (2010). What Is Sustainability Economics? In S. Bell, & S. Morse (Eds.), *Sustainability Indicators: Measuring the Immeasurable? Second* (2nd ed., Issue 69, pp. 445–450). Oxfordshire: Earthscan.
Benitez, K. (2018). *Social Justice and Sustainability Go Hand in Hand at Northwestern.* Northwestern University. https://www.northwestern.edu/sustainability/news/2018/2018-01-social-justice.html.
Bergek, A., Jacobsson, S., Carlsson, B., Lindmark, S., & Rickne, A. (2008). Analyzing the Functional Dynamics of Technological Innovation Systems: A Scheme of Analysis. *Research Policy, 37*(3), pp. 407–429. http://linkinghub.elsevier.com/retrieve/pii/S004873330700248X%5Cnhttp://urn.kb.se/resolve?urn=urn:nbn:se:liu:diva-46600.
Blowers, A. and Pain, K. (1999). The unsustainable city? In Pile, S., Brook, C., & Mooney, G. (Eds.), *Understanding cities: Unruly cities? Order/disorder* (p. 247). Milton Park, Abingdon-on-Thames, Oxfordshire: Routledge. ISBN 0415200733.
World Commission on Environment and Development. (1987). *The Brundtland Report: 'Our Common Future'.* (1st ed.). Oxford: Oxford University Press.
Climate Justice Alliance. (2022). *Just Transmission: A Framework for Change.* Accessed 29/03/2022 from https://Climatejusticealliance.Org/Just-Transition/.
Cohen, S. (2020, January 27). Economic Growth and Environmental Sustainability. *State of the Planet.* https://doi.org/10.4324/9780203011751.
Connors, R., Matthews, S., Chen, R. B. R., Zhou, Y., Gomis, X., Lonnoy, M. I., Maycock, E., & Tignor, T. (2018). *Sustainable Development, Poverty Eradication and*

Reducing Inequalities Book or Report Section Published Version. http://centaur.reading. ac.uk/81048/.

Echendu, A., & Georgeou, N. (2021). 'Not Going to Plan': Urban Planning, Flooding, and Sustainability in Port Harcourt City, Nigeria. *Urban Forum, 32*(3), pp. 311–332. https:// doi.org/10.1007/s12132-021-09420-0.

El Bilali, H. (2019). The Multi-Level Perspective in Research on Sustainability Transitions in Agriculture and Food Systems: A Systematic Review. *Agriculture (Switzerland), 9*(4). https://doi.org/10.3390/agriculture9040074a.

Elkington, J. (1997). *Cannibals with Forks: The Triple Bottom Line of the 21st Century Business.* Capstone.

Eswatini Government. (2007). *The Swaziland Poverty Reduction Strategy and Action Plan (PRSAP)* (Issue June). https://www.tralac.org/files/2012/12/Final-Poverty-Reduction-Strategy-and-Action-Plan-for-Swaziland.pdf.

Eswatini Government. (2019). *The Kingdom of Eswatini Voluntary National Review 2019 Report* (Issue June). https://sustainabledevelopment.un.org/content/ documents/24651Eswatini_VNR_Final_Report.pdf.

Eswatini Government. (2020). Kingdom of Eswatini Voluntary National Report. In *IMF Staff Country Reports* (Vol. 20, Issue 229). https://doi.org/10.5089/9781513551869.002.

Eswatini Ministry of Economic Planning and Development. (2017). *Strategy for sustainable development and inclusive growth (SSDIG) 2030.* Mbabane: Ministry of Economic Planning and Development.

European Bank. (2022). *Just Transition.* European Bank for Reconstruction and Development. https://www.ebrd.com/what-we-do/just-transition.

Figenbaum, E. (2017). Perspectives on Norway's Supercharged Electric Vehicle Policy. *Environmental Innovation and Societal Transitions, 25*, pp. 14–34. https://doi. org/10.1016/j.eist.2016.11.002.

Geels, F., & Raven, R. (2006). Non-linearity and Expectations in Niche-Development Trajectories: Ups and Downs in Dutch Biogas Development (1973–2003). *Technology Analysis and Strategic Management, 18*(3–4), pp. 375–392. https://doi. org/10.1080/09537320600777143.

Geels, F. W. (2014). Regime Resistance against Low-Carbon Transitions: Introducing Politics and Power into the Multi-Level Perspective. *Theory, Culture & Society, 31*(5), pp. 21–40. https://doi.org/10.1177/0263276414531627.

Government of Botswana. (2018). International Conference. *Leave No One Behind: The Fight against Poverty, Exclusion and Inequality*, Gaborone, Botswana, March, 1–17.

Government of Lesotho. (2019). The Kingdom of Lesotho voluntary national review on the implementation of the 2030 agenda report. Maseu. https://lesotho.un.org/en/36193-kingdom-lesotho-voluntary-national-review-implementation-2030-agenda-report-2019.

Government of Zimbabwe. (2018). Vision 2030 – Towards a prosperous and empowered upper middle income society by 2030. https://ucaz.org.zw/wp-content/uploads/2019/08/ Vision_2030_Republic_of_Zimbabwe_Web_Version.pdf.

Grin, J., Rotmans, J., & Schot, J. (2010). *Transitions to Sustainable Development: New Directions in the Study of Long Term Transformative Change.* London: Routledge.

Hammond, G. P. (2000). Energy, environment and sustainable development: A UK perspective. *Process Safety and Environmental Protection, 78*(4), 304-323.

Harding, T. J., Reemer, T. B., Coninx, I., de Rooij, L. L., Casu, F. A. M., Dijkshoorn-Dekker, M. W. C., Likoko, E. A., Eweg, A. Y., Koopmanschap, E. M. J., Mekonnen, D. A., & Termeer, E. E. W. (2022). *How to Operationalise Just Transitions?: Insights from Dialogues.* Wageningen: Wageningen Environmental Research.

Hielscher, S., Wittmayer, J. M., & Dańkowska, A. (2022). Social Movements in Energy Transitions: The Politics of Fossil Fuel Energy Pathways in the United Kingdom, the Netherlands and Poland. *The Extractive Industries and Society*, December 2021, p. 101073. https://doi.org/10.1016/j.exis.2022.101073.

International Renewable Energy Agency (ERENA). (2022). ERENA-AfDB Report: Energy Transition Centre to Africa's Economic Future. Retrieved 30/11/2022 from https://www.irena.org/news/pressreleases/2022/Jan/IRENA-AfDB-report.

IPPC. (2023). The Evidence Is Clear: The Time for Action Is Now. We Can Halve Emissions by 2030. https://www.ipcc.ch/2022/04/04/ipcc-ar6-wgiii-pressrelease/

Jenkins, K. E., Sovacool, B. K., Błachowicz, A., & Lauer, A. (2020). Politicising the Just Transition: Linking Global Climate Policy, Nationally Determined Contributions and Targeted Research Agendas. *Geoforum*, (115), pp. 138–142.

Johnstone, P., & Stirling, A. (2020). Comparing Nuclear Trajectories in Germany and the United Kingdom: From Regimes to Democracies in Sociotechnical Transitions and Discontinuities. *Energy Research and Social Science*, *59*(December 2018), p. 101245. https://doi.org/10.1016/j.erss.2019.101245.

Kemp, R., Schot, J., Hoogma, R., & Kemp, R., Shot, J., & Hoogma, R. (1998). Regime Shifts to Sustainability through Process of Niche Formation: The Approach of Strategic Niche Management. *Technology Analysis and Strategic Management*, *10*(2), p. 175.

Khavul, S., & Bruton, G. D. (2013). Harnessing Innovation for Change: Sustainability and Poverty in Developing Countries. *Journal of Management Studies*, *50*(2), pp. 285–306.

Kinn, M. C. (2016). *An Analysis of the Sociotechnical Transition Process From the Existing Centralised Alternating Current Voltage Electrical System in the Uk to One Where Distributed Direct Current Voltage Is Used to Meet the Energy Needs of the Built Environment* (Issue December 2016) [The University of Salford]. http://usir.salford.ac.uk/id/eprint/40962/3/Moshe Kinn PhD thesis 10–12–2016.pdf.

Köhler, J., Geels, F. W., Kern, F., Markard, J., Onsongo, E., Wieczorek, A., Alkemade, F., Avelino, F., Bergek, A., Boons, F., Fünfschilling, L., Hess, D., Holtz, G., Hyysalo, S., Jenkins, K., Kivimaa, P., Martiskainen, M., McMeekin, A., Mühlemeier, M. S., … Wells, P. (2019). An Agenda for Sustainability Transitions Research: State of the Art and Future Directions. *Environmental Innovation and Societal Transitions*, *31*(2019), pp. 1–32. https://doi.org/10.1016/j.eist.2019.01.004.

Leal Filho, W., Henrique Paulino Pires Eustachio, J., Dinis, M. A. P., Sharifi, A., Venkatesan, M., Donkor, F. K., & Vargas-Hernández, J. (2022). Transient Poverty in a Sustainable Development Context. *International Journal of Sustainable Development & World Ecology*, *1*(1), pp. 1–14.

Loorbach, D., Frantzeskaki, N., & Avelino, F. (2017). Sustainability Transitions Research: Transforming Science and Practice for Societal Change. *Annual Review of Environment and Resources*, *42*, pp. 599–626. https://doi.org/10.1146/annurev-environ-102014-021340.

Loorbach, D., & Rotmans, J. (2010). The Practice of Transition Management: Examples and Lessons from Four Distinct Cases. *Futures*, *42*(3), pp. 237–246. https://doi.org/10.1016/j.futures.2009.11.009.

Magnusson, T., & Werner, V. (2022). Conceptualisations of Incumbent Firms in Sustainability Transitions: Insights from Organisation Theory and a Systematic Literature Review. *Business Strategy and the Environment*, August 2020, pp. 1–17. https://doi.org/10.1002/bse.3081.

Markard, J., Geels, F. W., & Raven, R. (2020). Challenges in the Acceleration of Sustainability Transitions. *Environmental Research Letters*, *15*(8). https://doi.org/10.1088/1748-9326/ab9468.

Markard, J., Hekkert, M., & Jacobsson, S. (2015). The Technological Innovation Systems Framework: Response to Six Criticisms. *Environmental Innovation and Societal Transitions, 16*, pp. 76–86. https://doi.org/10.1016/j.eist.2015.07.006.

Markard, J., Raven, R., & Truffer, B. (2012). Sustainability Transitions: An Emerging Field of Research and Its Prospects. *Research Policy, 41*(6), pp. 955–967. https://doi.org/10.1016/j.respol.2012.02.013.

Martiskainen, M., & Sovacool, B. K. (2021). Mixed Feelings: A Review and Research Agenda for Emotions in Sustainability Transitions. *Environmental Innovation and Societal Transitions, 40*(October), pp. 609–624. https://doi.org/10.1016/j.eist.2021.10.023.

Mcneill, J. (2005). *The Employment Effects of Sustainable Development Policies* (Issue September 2014). https://doi.org/10.1016/j.ecolecon.2007.02.028

Parkin, S. (2000). Sustainable Development: The Concept and the Practical Challenge. *Proceedings of the Institution of Civil Engineers. Civil Engineering*, (138), pp. 3–8. http://doi.org/10.1680/cien.2000.138.6.3.

Pickett, S. T. A., Boone, C. G., Mcgrath, B. P., Cadenasso, M. L., Childers, D. L., Ogden, L. A., Mchale, M., & Grove, J. M. (2013). Ecological Science and Transformation to the Sustainable City. *Cities, 32*, S10–S20. https://doi.org/10.1016/j.cities.2013.02.008.

Pollin, R. (2021). *Achieving the Just Transition to a Sustainable Future*. SDG Action. https://sdg-action.org/achieving-the-just-transition-to-a-sustainable-future-2/.

Ramos-Mejía, M., Franco-Garcia, M. L., & Jauregui-Becker, J. M. (2018). Sustainability Transitions in the Developing World: Challenges of Socio-Technical Transformations Unfolding in Contexts of Poverty. *Environmental Science and Policy, 84*(April 2017), pp. 217–223. https://doi.org/10.1016/j.envsci.2017.03.010.

Ritchie, H., Roser, M., & Rosado, P. (2022). "Energy". Published online at OurWorldInData.org. Retrieved from https://ourworldindata.org/energy.

Romijn, H., Raven, R., & de Visser, I. (2010). Biomass Energy Experiments in Rural India: Insights from Learning-Based Development Approaches and Lessons for Strategic Niche Management. *Environmental Science & Policy, 13*(4), pp. 326–338. https://www.golder.com/insights/block-caving-a-viable-alternative/.

Roth, J., Dufour, L., & Gencsu, I. (2022). *Putting Poverty Alleviation and Inequality Reduction at the Heart of the Energy Transition*. https://www.iisd.org/articles/policy-analysis/poverty-inequality-energy.

Sabato, S., & Fronteddu, B. (2020). A Socially Just Transition through the European Green Deal? In *ETUI Research Paper-Working Paper*. https://doi.org/10.2139/ssrn.3699367.

Satgar, V. (2018). The Climate Crisis and Systemic Alternatives. In V. Satgar (Ed.), The Climate Crisis: South African and Global Democratic Eco-Socialist Alternatives (pp. 1–28). Johannesburg: Wits University Press. https://doi.org/10.18772/22018020541.6.

Schot, J., & Geels, F. W. (2008). Strategic niche management and sustainable innovation journeys: theory, findings, research agenda, and policy. *Technology Analysis & Strategic Management, 20*(5), 537–554.

Sharpe, S. A., & Martinez-fernandez, C. M. (2021). The Implications of Green Employment: Making a Just Transition in ASEAN. *Sustainability*, (13), pp. 1–19. https://www.iisd.org/articles/policy-analysis/poverty-inequality-energy.

Silva, A., & Stocker, L. (2018). What Is a Transition? Exploring Visual and Textual Definitions among Sustainability Transition Networks. *Global Environmental Change, 50*(June 2017), pp. 60–74. https://doi.org/10.1016/j.gloenvcha.2018.02.003.

Smith, A., Voß, J. P., & Grin, J. (2010). Innovation Studies and Sustainability Transitions: The Allure of the Multi-level Perspective and Its Challenges. *Research Policy, 39*(4), pp. 435–448. https://doi.org/10.1016/j.respol.2010.01.023.

Stevis, D., Morena, E., & Krause, D. (2022). Just Transitions: Social Justice in the Shift towards a Low-Carbon World. In E. Morena, D. Krause, & D. Stevis (Eds.), *Introduction: The genealogy and contemporary politics of just transitions* (1st ed., pp. 1–14). Pluto Press.

Svobodova, K., Owen, J. R., Harris, J., & Worden, S., 2020. Complexities and Contradictions in the Global Energy Transition: A Re-evaluation of Country-Level Factors and Dependencies. *Applied Energy*, *265*, p. 114778.

Swilling, M., Musango, J., & Wakeford, J. (2016). Developmental States and Sustainability Transitions: Prospects of a Just Transition in South Africa. *Journal of Environmental Policy and Planning*, *18*(5), pp. 650–672. https://doi.org/10.1080/1523908X.2015.1107716.

Thring, M. W. (1990). Engineering in a Stable World. *Science, Technology and Development*, *8*(2), pp. 107–121.

United Nations. (2022). *Poverty Eradication*. Poverty. https://www.un.org/development/desa/socialperspectiveondevelopment/issues/poverty-eradication.html.

United Nations – Eswatini. (2021). *A Prosperous, Just and Resilient Eswatini Where No One Is Left Behind*. UNDP (2023). SDG Investor Platform. Eswatini Summary Opportunities. Retrieved 07/02/2023 Eswatini | SDG Investor Platform (undp.org).

United Nations – Eswatini (2023). We Must Take Urgent Action to Reduce Inequalities. Retrieved 13/02/2023 from http://eswatini.un.org.

Weber, K. M., & Truffer, B. (2017). Moving Innovation Systems Research to the Next Level: Towards an Integrative Agenda. *Oxford Review of Economic Policy*, *33*(1), pp. 101–121. https://doi.org/10.1093/oxrep/grx002.

WFP. (2019). *Country Strategic Plan: Eswatini* (Issue November 2019). http://gender.manuals.wfp.org/en/gender-toolkit/gender-in-programming/gender-and-age-marker/.

World Bank. (2022). *Poverty and Equity Brief*. Eswatini: World Bank Group.

Yazdani, S., & Dola, K. (2013). Sustainable City Priorities in Global North versus Global South. *Journal of Sustainable Development*, *6*(7), pp. 38–47. https://doi.org/10.5539/jsd.v6n7p38.

Zhu, Y., Bashir, S., & Marie, M. (2022). Assessing the Relationship between Poverty and Economic Growth: Does Sustainable Development Goal Can Be Achieved? *Environmental Science and Pollution Research*, *29*(19), pp. 27613–27623. https://doi.org/10.1007/s11356-021-18240-5.

Zolfagharian, M., Walrave, B., Raven, R., & Romme, A. G. L. (2019). Studying transitions: Past, Present, and Future. *Research Policy*, *48*(9). https://doi.org/10.1016/j.respol.2019.04.012.

Ziervogel, G., Lennard, C., Midgley, G., New, M., Simpson, N. P., Trisos, C. H., & Zvobgo, L. (2022). Climate Change in South Africa: Risks and Opportunities for Climate-Resilient Development in the IPCC Sixth Assessment WGII Report. *South African Journal of Science*, *118*(9–10), pp. 1–5.

7 Climate Actions and Just Transition in Zimbabwe

A Review

Tendai Nciizah, Elinah Nciizah and Adornis D. Nciizah

Introduction

Climate change is not only an environmental issue as seen by its effects which has brought to the fore justice issues which require attention. People's livelihoods have been grossly affected, and the redressing of climate change through different climate actions such as climate adaptation and mitigation is further leading to an exacerbation of existing inequalities among poor people, particularly those who reside in less economically developing countries (LEDCs). This makes climate change to be one of the most global serious challenges derailing crucial development gains especially for African communities who strongly rely on rain-fed agriculture (Walsh et al. 2021). Riedy (2016) argues that climate change poses an existential threat to human civilisation. Although the impact of climate change is observed on a macro scale, it is the ordinary grassroots people who are facing the major negative effects of climate change due to their reliance on rain-fed agriculture and also their low adaptive capacity. For instance, two in every three Africans rely on subsistence farming for food and basic livelihoods (Walsh et al. 2021), and consequently, any disruption to the agricultural system has serious effects on agricultural productivity and food security.

The indicators of climate change include changes in the surface temperature, atmospheric water vapour, precipitation, severe weather events, glaciers, ocean and land ice and sea level (Cubasch et al. 2013). Other indicators include frequent occurrence of extreme weather events such as droughts, storms, floods and heat waves. The two major causes of climate change include natural causes and anthropogenic causes. However, in recent years, it is the human activities such as burning fossils like coal, oil and gas as well as unsustainable farming practices that are the main drivers of climate change. Climate change caused by human activities such as burning fossil fuels (coal, oil and natural gas) and clearing forests for farms and cities have increased in recent years increasing climate change. Stocker et al. (2013) point out that the Intergovernmental Panel on Climate Change (IPCC) fifth assessment report states that warming of the climate system is unequivocal, and that the atmosphere and ocean have warmed, sea level has arisen and greenhouse gas (GHG) emissions have increased. The key GHGs emitted by various human activities are carbon dioxide (CO_2), methane (CH_4) and nitrous oxide (N_2O). IPCC

DOI: 10.4324/9781003397120-9

(2014) adds more, which are ozone (O_3), water vapour (H_2O), hydro-fluorocarbons (HFC$_s$), per-fluorocarbons (PFC_2) and sulphur hexafluoride (SF_6). Notably is that carbon dioxide is the most GHGs produced by human activities accounting for about 76% of total anthropogenic gas emissions in 2010 (Moyo 2019). Also the use of fossil fuels and other industrial processes are by far the largest contributor of CO_2 followed by agricultural activities (11%) (IPCC 2014). Agricultural activities and biomass burning are the largest contributors of CH_4, while unsustainable use of nitrogen fertiliser and fossil-fuel combustion largely contribute to N_2O emissions. As of 2014, the burning of coal, natural gas and oil for electricity and heat is the largest single source of global GHG emissions (IPCC 2014). Statistics have shown that it is the developing countries that emit more emissions as compared to developing countries, yet it is the developing countries that are facing the major brunt of climate change. Global carbon dioxide emissions from fossil fuels and industry increased 5.3% in 2021 to reach a record high of 37.12 billion metric tonnes (GtCO$_2$) (Tiseo 2023). The two biggest contributors to global emissions that year were China and the United States, who produced 11.47 and 5.01 GtCO$_2$, respectively (Tiseo 2023).

Zimbabwe is emitting insignificantly like most developing countries but has to play a role in mitigation and adaptation so as to address the effects of climate change. It is important to mitigate climate change by reducing the flow of GHG emissions into the atmosphere by increasing the ability of sinks to store these gases (Fawzy et al. 2020).GHG sinks include soil, forests and oceans, and these sinks have a very high capacity for storing GHGs. There are various options for mitigating climate change such as using new and efficient technologies and renewable energies or changing management practices especially in agriculture and changing consumer behaviour (NASA 2022). However, in making these changes or transitions, it is important that socio-economic impacts are considered. For instance, before transitioning from a fossil-based industrial process to low-carbon one, it is important to consider the potential impact on workers. The need for such considerations gave rise to the just transition concept. This concept takes into consideration issues surrounding fairness in shifting to a low-carbon economy (Stockholm Environment Institute 2021). For a just transition to take place, there is a need for enabling policies as well as taking stock of all opportunities and potential challenges and solutions. History has shown that structural injustice has its roots in historical global inequalities following centuries of colonisation and exploitation. Inequality has continued to exist despite the independence of several developing countries which has made people from those countries to continue living in poverty with long histories of oppression. Climate change therefore adds a layer of vulnerability to the structural injustices that already affect developing countries (Williams et al. 2022). This therefore makes climate change a political, social and ethical issue which also needs to be understood from a capabilities approach. Climate justice theorists have articulated a range of frameworks for understanding the relationship between the effects of climate change and conceptions of justice and fairness. The capabilities approach can be applied as a climate justice theory which seeks to bring social and political recognition of specific and local vulnerabilities and the effects of climate change on the basic needs of human needs in various places and under different conditions (Schlosberg 2012). The importance of capabilities approach is that it brings to the fore the vulnerabilities that most people are faced with, that emanate from the effects of climate change and from the efforts meant to mitigate and adapt. A capabilities approach to climate justice is therefore significant as it can be used as a normative guideline for climate policies and offers concrete standards by which to measure progress (Schlosberg 2012). A just transition without integrating the capabilities-based approach will fail to address individual and community needs and vulnerabilities, therefore making the poor poorer and the rich richer. As a country, Zimbabwe is highly vulnerable to climate change as most of the people live in rural areas and in poverty. Zimbabwe has been exposed to climate risks and disasters and is therefore challenged to articulate climate justice issues in its climate change response strategies (Chanza et al. 2015). This chapter summarises efforts done by Zimbabwe in promoting a low-carbon economy with an essence to show if there is a promotion of just transition. This chapter calls for a capabilities approach to avoid impacting negatively on the ordinary people. It also addresses opportunities generated by a just transition and the potential challenges.

Impacts of Climate Change in Zimbabwe

Among the top ten countries most affected by climate change in 2019, five were African with Zimbabwe and Mozambique being in the top two (Walsh et al. 2021). The negative impacts of climate change have not spared Zimbabwe, with the impacts further exacerbated by the country's geographical position in the semi-arid belt of Southern Africa and also the country's reliance on rain-fed agriculture (GoZ 2016). The effects of climate change in the country include recurrent intense droughts, flooding, water scarcity, severe fires and declining biodiversity. Cyclone Idai that took place in 2019 was one of the catastrophic effects of climate change in Zimbabwe, affecting livelihoods and increasing the poverty rates in the country. Noteworthy is that the major effects of climate change in Zimbabwe are being felt adversely in the rural areas as there is heavy reliance on rain-fed agriculture. This makes the agricultural sector to be highly vulnerable to climate change impacts, especially variability and climate-induced natural hazards, yet agriculture is the backbone of the economy with 75% of the population relying on agriculture for their food. Many scholars have highlighted on the fact that in the early days of independence, Zimbabwe was considered the breadbasket of Africa as it provided food to other countries. This gives evidence to how agriculture has always been the main mainstay of the country's economy.

According to Mudimu (2003) and Poulton et al. (2002), in value terms, the agricultural sector (including commercial agriculture) has often been the largest single source of export earnings, contributing up to 40% of total exports at least before fast track. Although Zimbabwe's agricultural exports fell between 54% and 74% during the period from 2000 to 2007, largely due to low production and productivity arising from fast track, in 2010, agricultural exports contributed US$487.4 million to the US$1,253.4 million total national export proceeds. This represented a contribution of 38.9%, as against 22.9% for the fuels and mining sectors and 33.7% for manufacturing. The main agricultural exports included tobacco, sugar, beef, horticultural produce, coffee, tea and cotton lint, with tobacco and cotton accounting for 83% in 2009 (Anseeuw et al. 2012). Kapuya et al. (2010) thus point out that the contribution of agriculture, in terms of food production and cash crop exports, has been the pillar of Zimbabwe's economic stability, with years of drought often coinciding with negative economic growth rates. The importance of the agricultural sector is thus reflected in the strong and direct positive correlation between agricultural performance and overall economic growth. Annual Gross Domestic Product, for instance, declined during bad agricultural years (the droughts of 1983, 1987, 1992 and 1995) and increased during years of good agricultural performance (including 1996 and 2009) (Kapuya et al. 2010; Poulton et al. 2002). All of this is evidence of the importance of agriculture to the Zimbabwean economy and the Zimbabwean livelihoods. However, even though the agriculture system in the country was largely affected by the poor economic policies that affected commercial agriculture, currently, climate change is strongly affecting farmers.

The major impact of decline in agriculture due to climatic conditions has been food insecurity as climate change and variability destabilised food production in Zimbabwe. Brown et al. (2012) reiterate this by pointing out that agriculture's sensitivity to climate-induced water stress is likely to intensify the existing problems of declining agricultural outputs, declining economic productivity, poverty and food insecurity, with communal farmers particularly affected. In the 2011/2012 season, the country was forced to import over 50% of maize from other countries in order to meet the needs of its citizens (Manyeruke et al. 2013). Chikodzi et al. (2013) rightly argue that extreme weather events in Zimbabwe, notably droughts, floods and tropical storms, threaten any development gains pursued by the Zimbabwean government across a variety of sectors, as well as intensify existing natural hazard burdens for the vulnerable populations in both rural and urban areas. The occurrence of drought, in the past years, has had implications on the wider Zimbabwean economy as, for example, many people were laid off from their employment, which resulted in the erosion of income for the country (FAO 2004). Thus, the economic effects of the 1991/1992 drought were also felt outside the agricultural sector. Largely as a result of the drought, through water and electricity shortages, manufacturing output in Zimbabwe declined by 9.3% with a 25% reduction in volume of manufacturing output and a 6% decline in foreign currency receipts (Robinson 1993). Serious reductions in agricultural output resulted in reduced economic growth and the loss of the much-needed foreign exchange normally derived from agricultural exports (Anseeuw et al. 2012; CEEPA 2006; Chagutah 2010).

The predicament caused by drought in Zimbabwe is extreme, contributing to underdevelopment of the country in most respects. Drought has its greatest impact on water supplies. Lack of water affects every aspect of environmental health and human activity, including agriculture, natural areas and development projects. The 1991/1992 drought in Zimbabwe killed more than one million cattle. Drought leads to overgrazing which furthers the degradation of pastures and arable areas and thus reduces livestock numbers. The severe 1991/1992 drought in fact led to extensive commercial farm retrenchments, adding to Economic Structural Adjustment Programme-induced retrenchments, and increased pressure on arable land and natural resources in communal spaces (Moyo 2000). Numerous droughts continue to affect the country. In May 2007, the Famine Early Warning System Network (FEWS NET) reported that prolonged dry spells during the 2006–2007 harvest seasons in most southern districts had contributed to low yields of maize in particular. Protracted economic decline, exacerbated by the poor 2006–2007 harvest, caused a significant decrease in Zimbabwe's food security, especially in the south-west of the country and in urban areas (FEWS NET 2007).

The post-2013 period in Zimbabwe has been marked with droughts which took place during the 2014/2015 season and the 2015/2016 season. In 2014/2015, delayed and inadequate rains resulted in late planting and wilted crops (FAO 2016). The 2014/2015 cropping season was poor, such that households had lower maize reserves and had to rely more on markets to access food (FAO 2016). At that time, Zimbabwe's future appeared precariously poised on an edge as two consecutive years of poor rains, compounded by El Niño, have resulted in the worst drought in 35 years (Scoones 2015). With at least 70% of the population relying on agriculture for their livelihoods, the effects of El Niño are increasingly having a direct and major impact through loss of income from crop and livestock value chains, as well as alternative income-generating opportunities for vulnerable people in communal areas and elsewhere (FAO 2016). Drought conditions, in mainly the southern provinces of the country, have consequently had a negative impact on livelihoods and aggravated food insecurity (FAO 2016). Scoones (2015) stipulates that, by January 2016, at the height of the 2015/2016 lean season, 30% of the rural population was reported in need of humanitarian aid, and the country's global acute malnutrition (GAM) rate for children under five had reached 5.7%, its highest in 15 years. He adds that more than 23,000 head of livestock perished due to lack of water, pasture or drought-related disease. The government declared a disaster appealing for US$1.5 billion in national and international assistance (Scoones 2015). All this shows the negative impacts of climate change on the country and the repercussions it is having on the livelihoods of the people. The importance of understanding the impacts of climate change on economic growth and development cannot be undermined as it paves way for the designing of climate strategies meant to promote job creation, economic and social development, while minimising negative effects and leaving no one behind. Although the country emits a low percentage of GHG emissions, it is taking strides in shifting to a low-carbon economy as low emission does not equate to inaction. Mutasa (2019) points out that Zimbabwe's GHG emissions constitute about 0.062% of the global total, while the whole of Africa contributes less than 7%. The Government of Zimbabwe (GoZ 2012) highlighted that the sources of GHG emissions in Zimbabwe are energy (49%), agriculture (40%), waste (6%) and industry (5%). There is therefore a need for a just transition as the country shifts to a low-carbon economy.

Just Transition

Climate change requires people to change their ways of living through rigorous means which might be unfair to other people resulting in negative impacts, e.g. heightened poverty, unemployment and food insecurity. Yet at the same time, living with climate change and not doing anything about it will result in more painful impacts inclusive of worsening of global inequity, deepening of poverty and intense reduced agricultural yields. It is believed that just transition has the potential to address the social, economic and environmental justice concerns. Recent years have therefore witnessed a call for just transition in climate change with global treaties beginning to take into consideration the need for a just transition. According to Williams et al. (2022), just transition can be traced to labour unions and environmental justice groups who saw the need to remove industries that were bringing harm to workers, community health and the planet while at the same time providing just pathways for workers into new livelihoods. Hence, the core to a just transition is having deep democracy whereby workers and communities have control over the decisions that affect their daily lives.

The concept of just transition believes that the transformational change that is needed to address the drivers and the effects of climate change has the potential to negatively impact some actors while positively impacting others (African Development Bank Group 2022). As we move from carbon-intensive industries to low-carbon industries, workers will lose their jobs, leading to an increase in unemployment rates. According to Piggot et al. (2019), to ensure a vision of climate justice which is fair and equitable and works to address the multiple challenges we face, there is a need to promote a transformative just transition away from fossil fuels towards a more prosperous equitable renewable and climate-positive future or carbon-negative future where we were taking more carbon and GHG emissions out of the atmosphere than we are putting into it. According to the African Development Bank Group (2022), a just transition can enable African countries to achieve greater redistribution of wealth and ensure that climate action occurs alongside development. The essence is to remove industries that harm workers, community health and the planet, but at the same time, just transition is required so that people will get new jobs for survival. There is therefore the need to ensure that justice is implemented when carrying out policies meant to redress the effects of climate change.

According to Riedy (2016: 12), 'transition is a set of reinforcing changes in the interlocking technologies, practices, markets, institutions, infrastructure, cultures and values that make up society'. African Development Bank Group (2022) defines just transition as facilitating equitable access to the benefits and sharing of the costs of sustainable development such that the livelihoods of all people, including the most vulnerable, are supported and enhanced as societies to make the transition to low-carbon and resilient economies. A just transition in climate change therefore secures the future and livelihoods of workers and their communities in the transition to a low-carbon economy. Hence, the global climate change discourse is moving towards a greener world but without leaving traditional workers and their labour rights in the deep end (Chisaira 2019).

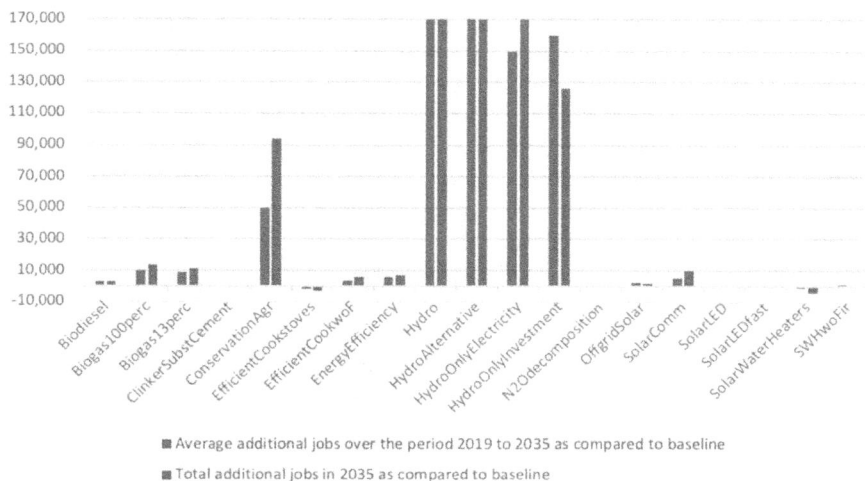

The chart shows two legend items:

■ Average additional jobs over the period 2019 to 2035 as compared to baseline
■ Total additional jobs in 2035 as compared to baseline

Figure 7.1 Framework for a Just Transition.

Figure 7.1 shows the strategy framework for a just transition. The essence is to move from an extractive economy that is informed by a colonial mind-set and a governance shaped by militarism to a regenerative economy that is shaped by a democratic governance. In order to do this, there is a need to promote a just transition as evidenced by social equity.

A Capabilities Approach to Climate Justice

The capabilities approach pioneered by Amartya Sen and further developed by Martha Nussbaum has been extensively used in the context of human development. Although Sen and Nussbaum offer different accounts of the capabilities approach with Sen focusing more on a development strategy that emphasises a variety of economic and social rights and Nussbaum focusing on how the approach can be used as a foundation for basic constitutional rights, they both incorporate justice-related concerns in their capabilities approach (Schlosberg 2012). These include distributional equity, social recognition and public participation (Schlosberg 2012), making the capabilities approach to be used for the evaluation and assessment of individual well-being, social arrangements and for the design of policies and proposals about social change in the society (Robeyns 2005). The approach argues that justice is mainly not about achieving an appropriate distribution of things between people, but it is about people being able to live lives that they consider worthwhile (Sen 1993).

According to Sen in Robeyns and Morten (2021), capabilities are the real freedoms that people have, in order to achieve their potential doings and beings. Sen and Nussbaum further state that the focus of social and economic development should be on wider human flourishing and on what people can achieve and do (Day et al. 2016), making the major core characteristic of the approach to be people's capabilities. By focusing on what people are able to do and be, rather than merely on the distribution of goods and resources, the capabilities approach recognises the diversity of people's ability to convert those resources and goods into real opportunities and achievements (Robeyns and Morten 2021). Hence, the approach functions as an overarching framework for explaining the interconnections between human well-being, the natural environment and technology (Hillerbrand 2015). It gives attention to the links that exist between material, mental and social well-being, i.e. economic, social and political and cultural dimensions (Robeyns 2005). The approach changes the focus from the resources that people have and the public goods they can access (means), to what they are able to do and be with those resources and goods (ends) (Robeyns and Morten 2021). By doing so, the capabilities approach becomes a people-centred approach which places human agency instead of organisations at the centre stage (Dreze and Sen 2002). Using this approach will lead to the acknowledgement that justice depends on a revised understanding of the relationship between human beings and the non-human world (Schlosberg 2015).

Considering that efforts to move to low-carbon economies entails a situation whereby some industries will be closed resulting in unemployment, avenues need to be created especially in countries that are taking strides to move to a low-carbon economy. Several people who will find themselves unemployed are already suffering due to poverty already in existence with climate change further reducing their capabilities. There is a need to use the capabilities approach in ensuring that climate adaptation and mitigation strategies do not negatively affect the livelihoods of the people. According to Amartya Sen, political opportunity to determine the capabilities necessary for our own functioning is central to a process of developing adaptation policies in response to local conditions and vulnerabilities. This is also integral to mitigation strategies that directly affect people's livelihoods. Considering that climate adaptation and mitigation strategies provide leeway for success, it is viable to implement the capabilities approach in order to avoid situations whereby these strategies may lead to increased inequalities. Hence, the approach becomes relevant in implementing climate justice particularly in developing countries such as Zimbabwe that grapple with inequalities and poverty. This is mainly because the approach has a global–local character as its realisation depends on specific local requirements making it applicable across political, economic and cultural borders (Wells 2012). According to Scholsberg (2012), Sen and Nussbaum point out that justice should not focus solely on distributive ideals; it should instead focus on a range of capacities necessary for people to develop free and productive lives that they design for themselves. Wood and Roelich (2019) supported this by stating that although reducing fossil-fuel combustion is necessary, there is a need to recognise that measures taken to reduce fossil-fuel combustion may have a more serious impact on those households that are already experiencing poverty. This is mainly because individuals differ greatly in their abilities of converting the same resources into valuable functionings. Using a capabilities approach in climate adaptation will therefore offer a way to assess vulnerability as it varies across location and scale (Scholsberg 2015). Wells (2012) pointed out that:

> from a justice perspective, the capability approach's relevance here is to argue that if people are falling short on a particular capability that has

been collectively agreed to be a significant one, then justice would require addressing the shortfall itself if at all possible, rather than offering compensation in some other form, such as increased income.

Hence, capabilities will have underlying requirements that vary strongly with different social circumstances. There are assets that people require in order to earn a living, and in a climate change scenario, these require attention and ways made possible for them to be acquired. In such cases, people's capabilities in light of the vulnerabilities they have need to be considered especially if a just transition is to be evidenced.

Just Transitions Efforts in Zimbabwe

There is very limited literature on just transition in Zimbabwe and very little has been done in terms of just transition in Zimbabwe. However, efforts are there as well as policies being drawn to promote just transition. Many countries are moving away from fossil fuels to renewable energy and are promoting low-carbon industries. According to United Nations Framework Convention on Climate Change (UNFCC) (2011), response to climate change by societies can come in the form of reducing GHG emissions and by adapting to its impacts. The report further points out that the capacity to adapt and mitigate is dependent on the socio-economic and environmental circumstances in a particular country and also on the availability of information and technology. This is also the case with Zimbabwe whereby the need to adapt and mitigate climate change depends on socio-economic aspects and investment meant to promote this. Through the Paris Agreement of 2015, Zimbabwe formulated the Nationally Determined Contribution (NDC), in order to mitigate climate change. The preamble of the Paris Agreement highlights the imperatives of a just transition of the workforce and the creation of decent work and quality jobs in accordance with nationally defined development priorities (UNFCCC 2015). Of paramount importance is that the Paris Agreement notes that the success of just transition is dependent on domestic climate policies of a country. Also for climate policies to be effective, they must be accompanied by just transition policies (UNDP Zimbabwe 2021). Hence, Zimbabwe is moving along with other countries in calling for the transformation to a low-carbon economy and has adopted its own National Climate Policy as a way of addressing this.

The National Climate Policy seeks to create a pathway towards a climate-resilient and low-carbon economy in which people have enough adaptive capacity to continue to develop in harmony with the environment (GoZ 2017). The National Climate Policy is supported by other policies which include the National Climate Change Response Strategy, Renewable Energy Policy, the National Environmental Policy, Biofuels Policy and the draft Forestry Policy (Zhakata 2019). Zhakata (2019) stipulates that the policy aims to guide climate change management in the country, enhance national adaptive capacity, increase mitigation actions and facilitate domestication of global policies and to ensure that there is compliance to the global mechanisms. In terms of low-carbon development, the policy sets out to develop low-carbon development pathways in the industrial, energy, waste, agriculture, land use, land-use change and forestry sectors (Zhakata 2019). Through Zimbabwe's NDC, the country has an obligation to reduce GHG emissions by 33% per capita below the projected business-as-usual scenario (GoZ 2015). Zimbabwe in their NDC report highlighted clean energy initiatives which include an increase in large and mini-hydropower plants, solar energy, ethanol blending, promotion of liquefied petroleum gas and construction of institutional biogas digesters and electrification of the rail system (Chisaira 2019). Major mitigation and adaptation strategies to be implemented in the country include implementation of renewable energy and energy efficiency initiatives, climate-smart agricultural practices, low-carbon transport systems, sustainable forest management, sustainable industrial development and solid waste management. The country is also aiming at searching for sustainable energy alternatives to the curing of tobacco (GoZ 2015).

The NDC has been revised in order to encompass new developments of implementing just transition in its low-carbon development pathways. This is because the move to a low-carbon economy will affect traditional fossil-fuel-based industries and will lead to the retrenchment of workers who were employed in these industries. The country should therefore encompass just transition in all its climate change policies in order to cater for all the workers who will be affected. A capabilities approach is also required to ensure that vulnerabilities of the workers are identified and support is given in such transitions. According to United Nations Development Programme (UNDP), Zimbabwe (2021), Zimbabwe's National Development Strategy 2021–2025 is the first economic blueprint on a trajectory of becoming a prosperous and empowered upper middle-income society by 2013, and in order to achieve this, new climate sensitive jobs amounting to at least 760,000 need to be created. Chisaira (2019) argues that even though the move to low-carbon economy will lead to retrenchment of workers, it will also lead to more jobs which will become available in the green economy sectors such as biofuel production, renewable energy plants, green housing construction, water management and resilience-building sectors.

The country also carried out a Green Jobs Assessment as a move to prepare for a low-carbon economy carried out alongside a just transition. To be noted is that Zimbabwe is among the first countries in the world to take a comprehensive review of its NDCs supported by just transition dialogue and Green Job Assessments Model (GJAM). The GJAM is an input–output model with an economic core, and it combines a macroeconomic model that is solved iteratively. According to the GoZ in its Zimbabwe Long-term Low-Greenhouse Gas Emission Development Strategy (2020–2050) Report, there is a need to make sure that transition to a low-carbon economy Zimbabwe will be just and inclusive transition for all. The phasing out of carbon-intensive industries and the ushering in of cleaner industries should sustain growth and employment, with both positive and negative effects on jobs and livelihoods put into consideration (Zimbabwe Long-term Low-Greenhouse Gas Emission Development Strategy [2020–2050] Report). Transition towards a low-carbon and sustainable economy will result in negative and positive impacts on employment. Putting this into consideration, the Zimbabwe's Low-Emission Development Strategy (LEDS) seeks to transform the economy towards a low-emission pathway, anchored on the creation of sustainable green jobs (Zhakata 2019). The LEDS is in line with the vision of the country to become an upper middle-income economy by 2030. In Zimbabwe, by participating in Conference of the Parties delegations, the Zimbabwe Congress of Trade Union (ZCTU) and the Labour and Economic Development Research Institute are playing a role in the inclusion of labour interest in climate change decision-making (Chisaira 2019). The country is calling for climate-smart and conservation framing as adaptation methods that will reduce the effects of climate change. According to UNDP Zimbabwe (2021), this move will entail several significant labour market effects as it requires an increase in organic fertiliser use and production. This will create jobs in supplying industries and calls for some 10% additional direct agriculture-related jobs in soil preparation, management, harvesting and post-harvest activities (UNDP Zimbabwe 2021). The country is also planning to move to biogas and improved cook stoves, and this will reduce emissions; however, social protection is required for any income losses on those households that depend on firewood, notably women (UNDP Zimbabwe 2021). The government also has plans to build two large-scale hydro power plants, namely the Bakota Gorge and the Devil's Gorge, which will lead to more job creation (UNDP Zimbabwe 2021). It is therefore crucial for the government to invest more in programmes that will create more jobs in order to make sure that workers are not affected. The economic decline in the country has already left people vulnerable and languishing in poverty which calls to fore the need to look into the capabilities that people require in order to survive in a world where countries are promoting mitigation strategies.

In as much as the country has outlined how it intends to shift to a low-carbon economy, there are some contradictions looking at some of the development policies that came with the new dispensation under President E. Mnangagwa. One of the development policies under the new dispensation is to attract investment in the extraction of coal and coal-bed methane in Hwange and Lupane districts (Chidarara 2019). Ironically, in 2017 the country drafted a policy for oil and gas extraction known as the Zimbabwe Oil and Gas Industry Development Policy at a time that the world is now calling for a reduction in carbon emissions as a way to mitigate climate change. Such a scenario taking place in the country shows how far the country's efforts to introduce a low-carbon economy are. According to Chidarara (2019), the country intends to expand coal production at a time when other countries are downscaling investments in coal. Considering this, it is therefore worrisome and questionable if the country will be able to move to a low-carbon economy and also implement just transition. The country is also suffering from economic decline with most of its citizens unemployed and food insecure. Such a situation makes it difficult to successfully shift away to a low-carbon economy. There is no way that just transition can happen without including local citizens, communities and businesses, and Zimbabwe is still far away from this. Walsh et al. (2021) postulate that as people's livelihoods come under increasing pressure, they are often left with little option other than to exploit resources unsustainably, such as engaging in illegal logging for charcoal production or by using high-carbon energy sources. This is the situation that is taking place in Zimbabwe with a rise in cutting down trees and the use of charcoal as a fuel for cooking. According to UNDP Zimbabwe (2021) deforestation, burning of biomass and agriculture are the largest sources of Zimbabwe's emissions. The constant power cuts in urban areas and the lack of rural electrification in most of the rural areas makes the large population to depend on firewood as a cheaper means of fuel for cooking. This has increased the rate of deforestation in the country, and with the current outlook, the situation is far from ending. The country still heavily relies on coal-powered thermal stations that emit large amounts of pollution. Zimbabwe is also currently experiencing a high influx of small-scale artisanal miners who are mining unsustainably. In addition, there are countries such as China who are exploiting the fossil fuels in the country making mitigation difficult and just transition next to impossibility. This is likely to predict that it is going to be difficult for Zimbabwe to move from an extractive economy to a regenerative economy.

Opportunities and Challenges for a Just Transition in Zimbabwe

Opportunities

For countries to realise the full benefits of transitioning to a low-carbon economy, there is a need for enabling policies. In Zimbabwe, several climate change policies have been drafted to enable migration to a low-carbon economy through several projects, e.g. dam construction and rehabilitation for hydro-electricity, solar energy, biogas, etc. These projects will create thousands of job opportunities upon completion. For instance, upon completion, the mega-multi-billion-dollar Batoka dam will add approximately 170,000 jobs, while other projects such as commercial solar will add another 10,000 jobs in 2035 (UNDP Zimbabwe 2021). Figure 7.2

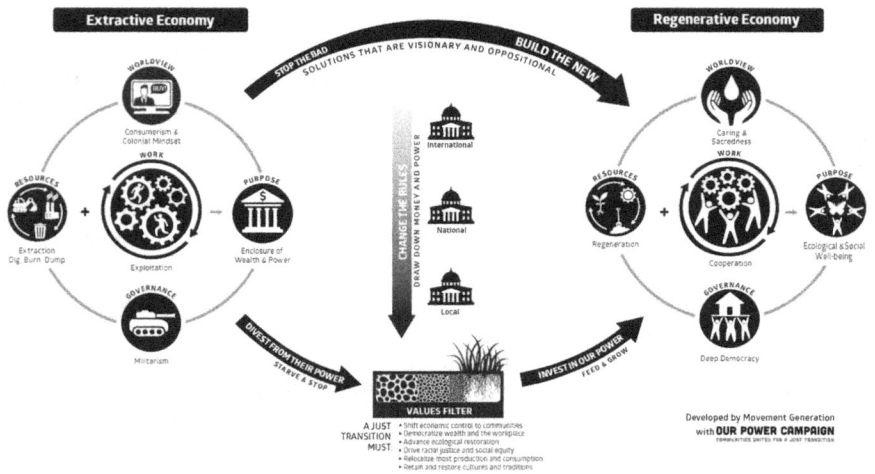

Figure 7.2 Additional Jobs.

summarises the total number of jobs that can be created under various projects (UNDP Zimbabwe 2021).

It has been indicated that strategies on fiscal, macroeconomic, sectoral and industry policies have the potential to support structural economic change and enhance economic growth and social development. According to UNDP Zimbabwe (2021), a main instrument is a fiscally neutral policy reform with a double dividend; tax carbon while lowering labour costs. This could shift economic growth to low-carbon activities and industries and, simultaneously, reduce the cost of employment, thereby enhancing overall national employment creation. Such efforts show that Zimbabwe is somehow making efforts to promote just transition by also observing the rights of workers and trying to create new opportunities for new employment. Chisaira (2019) postulates that the new clean energy initiatives pose opportunities for Zimbabwe's energy workforce if workers are re-trained; however, the challenge will be in the acquiring of finance.

The transition to a low-carbon economy has the potential to improve the quality of life through better working conditions and income (Wang et al. 2018). The use of improved and more efficient processes in the workplace provides a safe and healthy environment for workers compared to previous technologies. Moreover, the use of clean energy and sustainable production methods in the agricultural sector will not only provide a safe environment for workers but will also provide safe and healthy produce to consumers.

Challenges

There are many challenges that are associated with reducing the effects of climate change which in turn affect the implementation of the concept of just transition. The necessity for reductions in fossil-fuel consumption presents major risks to African countries that are beginning to look for the exploitation of carbon deposits in order to drive their development, just as the developed countries have done (Walsh et al. 2021). There is the need for the world to phase out carbon emissions, but other countries, especially those with advanced economies, have already benefitted from fossil-fuel resources (Walsh et al. 2021). The LEDCs, who are yet to benefit from fossil-fuel resources, are now being compelled to transition to a low-carbon economy. Most of these countries are still suffering from a plethora of problems and are still not fully equipped to shift to an economy that does not rely on fossil fuels. For instance, as mentioned above, most Zimbabweans rely on natural resources for their daily survival and are not yet equipped to have another livelihood that is divorced from natural resources. There is a need for the provision of alternative livelihoods to the people by the government in order to have a just transition that will benefit everyone. In order to make sure that the alternative livelihoods provided for the people actually cater for the people's needs, a capabilities approach needs to be implemented as a way of making sure that justice prevails in both the adaptation and mitigation strategies. Where climate-smart agriculture is being promoted, there is a need to provide the required technologies that will promote these and assist the people to transition. Failing to look into such capabilities will make the people suffer, and adaptation and mitigation efforts will fail.

Moving to a low-carbon economy will also result in companies using older technologies scaling down which will result in job losses. For instance, to reduce reliance on fossil energy in electricity production, coal-powered stations need to be closed. This will result in significant job losses. To offset this, there is a need for the government and other stakeholders to ensure the creation of sufficient green jobs that will benefit the affected local people. Putting all this into consideration, it is important to integrate the capabilities approach as it seeks to involve everyone from the local level to the national level. Sen (2009) argues for capabilities to be the basic currency by which questions of justice are decided. The capabilities approach makes the grassroots people get involved which promotes climate justice as it draws information on the capabilities that people have or need so as to live a dignified life. The capabilities approach seeks to know what people value, what they possess, the resources that they have and the issues that affect them in order to survive. In so doing, it exposes their vulnerabilities which is integral to understanding climate justice efforts. For instance, to transition to a regenerative economy, Zimbabweans have to be involved so that they will not lose their livelihoods as most depend on natural resources which make them embark, for instance, on illegal mining activities.

It has also been realised that well-intended climate policies and capital investments in the low-carbon economy require that managers, workers, enterprises and entrepreneurs have the right skills to finance, manage, construct, operate and maintain the capital asset and use it productively in the long term (UNDP Zimbabwe 2021). A lack of skilled managers and workers may hinder climate projects and investments; a skills development strategy should thus be developed for each priority sector. This is a huge challenge as it requires a lot of investment for this to be achieved. For instance, it requires integrating the required skills and professions into the country's education and training systems. It also involves creating a mechanism by which government, employers and workers discuss and decide on skills requirements, develop and update curricula, train teachers and trainers, integrate curricula in schools, technical vocational education and training (TVET) institutions and universities and roll out the skills strategy (UNDP Zimbabwe 2021). All this requires huge capital to train people to new jobs and is thus a major challenge in a country such as Zimbabwe that is already facing development challenges.

Conclusion

In conclusion, this chapter first highlighted the presence of climate change in Zimbabwe and its impact on agriculture and the country's economy. This chapter highlighted on a need to move away from extractive economies to low-carbon economies as countries aim to reduce GHG emissions. There is a need for a just transition in all climate actions such as climate adaptation and mitigation strategies in order to make sure that people's livelihoods do not continue being affected. In order to attain climate justice for the Zimbabwean people, this chapter called for the use of the capabilities approach. The capabilities approach will ensure that people's vulnerabilities that emanate from the effects of climate change and from the efforts meant to mitigate and adapt are brought to the forefront. This chapter further highlighted on the efforts being made in Zimbabwe in promoting just transition by drafting policies to be implemented in different sectors of the economy with the hope for great transformation by 2035. Zimbabwe's LEDS seeks to transform the economy towards a low-emission pathway, anchored on the creation of sustainable green jobs. In most respects, Zimbabwe is in its initial stages of working towards a just transition, hoping to have achievements by 2035. There is thus a need for dedication and commitment in order to achieve just transition. There are also other challenges that will affect a just transition in the country. For instance, in the case of Zimbabwe, the need to mitigate climate change while observing just transition depends on socio-economic aspects, availability of funds and investment meant to promote this. If the process of promoting just transition is done properly, there are, however, opportunities that can be evidenced which include the creation of new employment and improved quality of life.

References

African Development Bank Group, 2022, *Just Transition Initiative to Address Climate Change in the African Context*. African Development Bank.

Anseeuw, W., Kapuya, T., Saruchera, D., 2012, *Zimbabwe's Agricultural Reconstruction: Present State, Ongoing Projects and Prospects for Reinvestment*, Development Planning Division, Working Paper Series No. 32.

Brown, D., Chanakira, R. R., Chatiza, K., Dhliwayo, M., Dodman, D., Masiiwa, M., Muchadenyika, D., Mugabe, P., Zvigadza, S., 2012, *Climate Change Impacts, Vulnerability and Adaptation in Zimbabwe*, IIED, Working Paper No. 3.

CEEPA, 2006, *Climate change and African agriculture policy note No. 10. Centre for Environmental Economics and Policy in Africa (CEEPA)*, Pretoria: University of Pretoria.

Chagutah, T., 2010, *Climate Change Vulnerability and Preparedness in Southern Africa: Zimbabwe Country Report*. Heinrich Boell Stiftung.

Chanza, N., Chirisa, I., Makura, E. (2015). Ethical and Justice Reflections in Zimbabwe's INDC and Climate Policies. https://doi.org/10.13140/RG.2.1.2205.6561.

Chidarara, D., 2019, Climate Change, the Mining Industry and the law in Zimbabwe. In *Climate Change Law in Zimbabwe: Concepts and Insights* (eds): Murombo, T., Dhliwayo, M., Dhlakama, T. Konrad. Adenauer Foundation, pp. 139–164.

Chikodzi, D., Zinhiva, H., Simba, F. M., Murwendo, T., 2013, Reclassification of Agro-ecological Zones in Zimbabwe – The Rationale, Methods and Expected Benefits: The Case of Masvingo, *Journal of Sustainable Development in Africa*, Vol. 15, No. 1, pp. 104–116.

Chisaira, L. T., 2019, Climate Change and Labour Law in Zimbabwe: A Critical Perspective. In *Climate Change Law in Zimbabwe: Concepts and Insights* (eds): Murombo, T., Dhliwayo, M., Dhlakama, T. Konrad Adenauer Foundation, pp. 165–179.

Cubasch, U., D. Wuebbles, D. Chen, M.C. Facchini, D. Frame, N. Mahowald, and J.-G. Winther, 2013: Introduction. In: *Climate Change 2013: The Physical Science Basis. Contribution of Working Group I to the Fifth Assessment Report of the Intergovernmental Panel on Climate Change* (eds): Stocker, T.F., Qin, D., Plattner, G.-K., Tignor, M., Allen, S.K., Boschung, J., Nauels, A., Xia, Y., Bex, V., Midgley, P.M. Cambridge, New York: Cambridge University Press, pp. 119–144.

Day, R., Walker, G., Simcock, C., 2016, Conceptualising Energy Use and Poverty Using a Capabilities Framework, *Energy Policy*, Vol 93, pp. 255–264.

Dreze, J., Sen, A., 2002, *India: Development and Participation*. Oxford University Press.

FAO, 2004, *Drought and Climate Variability in the Limpopo River Basin*. FAO Corporate Document Repository.

FAO, 2016, *Southern Africa El Nino Response Plan 2016/17*. Rome: FAO.

FEWS (Famine Early Warning Systems Network), (2007), *Zimbabwe Food Security Emergency: Economic Decline, Poor Harvest Cause Significant Decrease in Food Security*. FEWS. Available at: http://v4.fews.net/docs/Publications/1001381.pdf.

Government of Zimbabwe (GoZ), 2012, *Zimbabwe's Second National Communication to the UNFCCC*. Harare: Ministry of Environment and Natural Resources Management.

Government of Zimbabwe (GoZ), 2015, *Zimbabwe's Intended Nationally Determined Contribution*. Harare: Ministry of Environment, Water and Climate.

Government of Zimbabwe (GoZ), 2016, *Zimbabwe National Climate Policy 2016*. Harare: Ministry of Environment, Water and Climate.

Government of Zimbabwe (GoZ), 2017, *Zimbabwe's National Climate Policy*. Harare: Ministry of Environment, Water and Climate.

Fawzy, S., Osman, A.I., Doran, J., Rooney, D.W., 2020, Strategies for Mitigation of Climate Change: A Review, *Environmental Chemistry Letters*, Vol. 18, No. 2020, 2069–2094, https://doi.org/10.1007/s10311-020-01059-w

Hillerbrand, R., 2015, The Role of Nuclear Energy in the Future energy Landscape: In *The Ethics of Nuclear Energy: Risk, Justice and Democracy in the Post-Fukushima Era* (eds): Taebi, B., Roeser, S., Cambridge: Cambridge University Press, pp. 231–249.

Intergovernmental Panel on Climate Change, (2014), AR5 Synthesis Report: Climate Change 2014. Available at: https://www.ipcc.ch/report/ar5/syr/

Kapuya, T., Saruchera, D., Jongwe, A., Mucheri, T., Mujeyi, K., Ndobongo, L. T., Meyer, F. H., 2010. *The Grain Industry Value Chain in Zimbabwe*. Unpublished Draft Prepared for the Food and Agricultural Organization (FAO). Available at www.fao.org/fileadmin/templates/est/AAACP/eastafrica/UnvPretoria_GrainChainZimbabwe_2010_1_pdf

Manyeruke, C., Hamauswa, S., Mhandara, L., 2013, The effects of Climate Change and Variability on Food Security in Zimbabwe: A Socio-Economic and Political Analysis. *International Journal of Humanities and Social Science*, Vol. 3, No. 6, pp. 270–286.

Mutasa, C., 2019, Zimbabwe's Climate: Past, Present and Future Trends. In *Climate Change Law in Zimbabwe: Concepts and Insights* (eds): Murombo, T., Dhliwayo, M., Dhlakama, T. Harare: Konrad. Adenauer Foundation, pp. 11–47.

Moyo, S., 2000, *Land Reform under Structural Adjustment in Zimbabwe*. Land Use Change in Mashonaland Provinces, Uppsala: Nordiska Afrika Institutet.

Mudimu, G., 2003, *Zimbabwe Food Security Issues Paper. Forum for Food Security in Southern Africa*. Overseas Development Institute (ODI). Available at: www.odi.org.uk/projects/03-food-security-forum/docs/ZimbabweCIPfinal.pdf.

Poulton C, Davies R, Matshe I, Urey I, 2002, *A review of Zimbabwe's Agricultural Economic Policies: 1980-2000*, ADU Working Papers 10922, Imperial College at Wye, Department of Agricultural Sciences.

Riedy C, 2016, Climate Change, Institute for Sustainable features, University of Technology, Sydney

Robeyns, I., 2005, The Capability Approach: A Theoretical Survey, *Journal of Human Development* Vol. 6, No. 1, pp. 93–117.

Robeyns, I., Morten, F. B., 2021, *The Capability Approach, The Stanford Encyclopaedia of Philosophy* (ed.): Edward N. Zalta. Available at https://plato.stanford.edu/archives/win2021/entries/capability-approach/.

Robinson, P., 1993, *Economic Effects of the 1992 Drought on the Manufacturing Sector in Zimbabwe*, London: Overseas Development Institute.

Sen, A., 1993, Capability and Wellbeing. In *The Quality of Life* (eds.): M. Nussbaum and A. Sen, Oxford: Clarendon Press, pp. 30–53. Available at https://doi.org/10.1093/01982 87976.003.0003.

Sen, A., 2009, *The Idea of Justice*. London: Allen Lane.

Scoones, I., 2015, *Sustainable Rural Livelihoods and Rural Development*. Rugby: Practical Action Publishing.

Schlosberg, D., 2012, Climate Justice and Capabilities: A Framework for Adaptation Policy. *Ethics and International Affairs,* Vol. 26. Available at https://doi.org.10.1017/S0892679412000615.

Stocker, T. F., Plattner, G. K., Tignor, M., Allen, S. K., Boschung, J., Nauels, A., Xia, Y., Bex, V., Midgley, P. M. (eds), 2013, *Climate Change 2013: The Physical Science Basis*. Contribution of Working Group 1 to the Fifth Assessment Report of the Intergovernmental Panel on Climate Change. Cambridge: Cambridge University Press.

Stockholm Environment Institute, 2021, *Leading a Green Recovery: Annual Report 2020*, Stockholm: The Stockholm Environment Institute.

Piggot, G., Boyland, M., Down, A., Torre, A.R., 2019, *Realizing a Just and Equitable Transition Away from Fossil Fuels*, Stockholm: The Stockholm Environment Institute.

Tiseo, I., 2023, *Annual Global Emissions of Carbon Dioxide 1940–2021*. Statista.

UNDP, 2021, *Zimbabwe Green Jobs Assessment Report: Measuring the Socioeconomic Impacts of Climate Policies to Guide NDC Enhancement and a Just Transition*. UNDP, NDC Support Programme, ILO.

UNFCCC, 2011, Durban Climate Change Conference. Available at http://unfccc.int/meetings/durban_nov_2011/meeting/6245.php.

UNFCCC, 2015, Adoption of the Paris Agreement. Available at http://unfccc.int/resource/docs/2015/cop21/eng/l09r01.pdf.

Wang, Z., Danish, Zhang, B., Wang, B., 2018, Renewable Energy Consumption, Economic Growth and Human Development Index in Pakistan: Evidence form Simultaneous Equation Model. *Journal of Cleaner Production*, Vol. 184, pp. 1081–1090. Available at https://doi.org/10.1016/j.jclepro.2018.02.260.

Walsh, G., Ahmed, I., Said, J., Maya, M. F., 2021, *A Just Transition for Africa: Championing a Fair and Prosperous Pathway to Net Zero*. Tony Blair Institute for Global Change.

Wells T.R, 2012, Sen's Capability Approach, Internet Encyclopaedia of Philosophy. Available at https://iep.utm.edu/.

Williams, J., Chin-Yee, S., Maslin, M., 2022, *Africa and Climate Justice at COP27 and Beyond: Impacts and Solutions Through an Interdisciplinary Lens*. UCL Open: Environment Preprint. DOI:10.14324/111.444/000180.v1

Woody, N., Roelich, K., 2019, Tensions, Capabilities and Justice in Climate Change Mitigation of Fossil Fuels, *Energy Research and Social Science*, Vol. 52, pp. 114–122.

Zhakata, W., 2019, Governing Climate Change: General Principles and the Paris Agreement: In *Climate Change Law in Zimbabwe: Concepts and Insights* (eds): Murombo, T., Dhliwayo, M., Dhlakama, T. Harare: Konrad Adenauer Foundation pp. 48–60.

8 Climate Governance, Inaction and Injustice in Buffalo City, South Africa

Philani Moyo

Introduction

It is widely accepted that climate governance broadly involves climate negotiations, policymaking, marshalling institutions and processes towards collective climate policy implementation and action. This continuous process happens through networked governance at different scales, from the sub-national to national and international (Coen *et al.* 2020) "involving a diverse group of national and local governments, international organisations, the private sector, non-governmental organisations and other social actors. Its purpose is to promote opportunities and prompt action to address climate change" (González and Numer 2020: 10). Accordingly, the collective aim of climate governance is to solve globally shared problems stemming from the climate emergency. While climate actions for responding to the climate emergency are done at the local and national level, they are collectively responding to a global problem.

The global climate governance system is primarily mediated by the United Nations Framework Convention on Climate Change (UNFCCC). A litany of geostrategic considerations, neoliberal economic interests and transnational relations dominated by the Global North drive the UNFCCC agenda and nature of climate governance. That is to say their economic and political interests, their environmental and climate lobby networks as well as non-state actors actively steer the direction of climate governance at that transnational level (Andonova *et al.* 2009). For these, and other, reasons, the transnational UNFCCC regime has been widely criticised (Jordan *et al.* 2015) for failing to equitably spearhead achievement of emission reduction targets (Intergovernmental Panel on Climate Change 2022), for its inability to lower global temperature increase to less than 1.5°C above the pre-industrial average (Atwoli *et al.* 2021; Lindsey and Dahlman 2022) and failure to raise US\$100 billion/annum climate finance for supporting climate action in the Global South (United Nations 2015). This robust criticism should not, however, be limited to the UNFCCC-led international climate governance system. It should be extended to the omissions of individual countries, including those in the Global South such as South Africa, as they are constituent parties of the transnational climate governance architecture.

DOI: 10.4324/9781003397120-10

South Africa's climate governance system and processes cascade from the international UNFCCC climate regime to its national policies right down to provincial strategies and local government-level climate action plans. The country's climate policy, strategies and action plans are built on principles of various pieces of legislation which are the legal component of its climate governance. The national constitution is the primary legal foundation, specifically Chapter 2, Section 24 a, b (i, ii, iii) on environmental rights, conservation and protection (Government of South Africa [GovSA] 1996) with supporting legislation that includes the National Environmental Management Biodiversity Act 10 of 2004 (GovSA 2004b), Disaster Management Amendment Act of 2002 (GovSA 2002), National Energy Act of 2008 (GovSA 2008a), National Environmental Management: Integrated Coastal Management Act No. 24 of 2008 (GovSA 2008b) and Carbon Tax Act No. 15 of 2019 (GovSA 2019a). Among others, these statutes, in their normative nature, simply provide a legal framework for national, provincial and local governments to invest resources in environmental interventions and climate action without prescribing the nature or form of such.

As part of this climate governance legal continuum, and as per its governance mandate, the South African government has over the years adopted a number of macro climate policies, strategies and micro plans that are the basis of its monocentric top-down climate governance across its three spheres of the government (national, provincial and local). These include, but not limited to, the National Development Plan (NDP) 2030, National Climate Change Response Strategy (NCCRS), National Climate Change Response White Paper (NCCRP), Long-Term Mitigation Strategy (LTMS), National Climate Change Adaptation Strategy (NCCAS), Green Transport Strategy (GTS), Integrated Resource Plan (IRP), Low-Emission Development Strategy (LEDS) and the Just Transition Framework (JTF). A critical examination of these in so far as they relate to climate governance actions is one of the subjects of discussion hereunder. For now, it suffices to note that these national policies and strategies have not been fully implemented by any measure. For example, while the strategic objectives of the NCCRS and NCCAS are clear, there is consistent failure by the national government to translate these into funded programmes at provincial and local levels. This absence of practical and structured implementation is glaring evidence of climate governance failure at the national level. Since the national government is failing to effectively respond to the climate emergency, what are the chances of a city (in this case Buffalo City in the Eastern Cape Province) to use its tangible and intangible resources to implement sustainable and impactful climate actions within the context of a multilevel climate governance system? To what extent does the polity and politics of a three-tier system of government in South Africa present and perpetuate climate action barriers in Buffalo City leading to climate injustice? What opportunities are presented by the growing polycentric system of climate governance as a pathway towards climate justice? These are the three primary questions answered by this chapter. In so doing, this chapter contributes to the evolving climate governance literature in three ways. First, it analyses and exposes climate action deficiencies and failures of a monocentric top-down climate governance regime in urban spaces. Secondly, it lays bare the influence of city polity and politics in perpetuating climate injustice. Lastly, it expands arguments on the polycentric approach in climate governance through an exposition of the growing influence of non-state actors in climate action in an urban space.

Climate Governance in South Africa: Revisiting Selected National Policies and Strategies

At the national level, the NDP 2030 makes a good policy case for climate action as it notes the existential and human development challenges posed by the climate emergency. Its mitigation agenda is premised on protection of the natural environment, diversification of energy sources towards sustainable renewable energy, entrenchment of a carbon pricing mechanism by 2030, investment in new climate-smart agricultural technologies, *ex ante* improved climate disaster preparedness, increased climate finance and a just transition (GovSA 2012). However, its mitigation objective is stymied by the need to set the country on a pathway towards a low-carbon future without compromising current fossil fuel dependent economic growth, jobs and livelihoods. This is due to the fact that no matter how one tries to interpret and frame the NDP's vision of a transition to a low-carbon climate-resilient green economy, it is inescapable that a large part of the South African economy, especially the energy-industrial complex, is dependent on coal and fossil fuels which are high greenhouse gas (GHG) emitters. Similarly, the country's motor and aviation industry are also driven by fossil fuels usage without restraint. Given this high-carbon energy-industrial structure of the economy and the millions of people dependent on it for jobs and livelihoods, it is obvious that South Africa requires a longer timeframe to practically transition to a low-carbon future without choking. The projected 2030 mitigation milestones of the NDP blueprint and the country's international commitments under the Paris Agreement's Nationally Determined Contributions (NDCs) are impossible to attain within current time-frames and under the prevailing neoliberal-driven processes of production and accumulation.

Another weakness of the NDP's climate governance agenda is that it openly ignores fundamental historical linkages between the climate emergency, neoliberal capitalism and causative role of the Global North in this matrix. While it talks about global partnerships that will propel the country to "manage the transition to a low-carbon economy" (GovSA 2012: 48), it places the primary burden of mitigation and adaptation action on the South African government and, to some extent, the local private sector. Prime GHG emitters and polluters of the Global North are exonerated from this primary responsibility as their role is reduced to that of 'global partners'. This places the NDP at odds with the global cardinal 'Polluter Pays Principle' (PPP) when dealing with global common problems such as the changed climate. Khan (2015: 638) notes that this variance is puzzling since the PPP is an "economic and ethical instrument" with "potential of effecting global responsibility for adaptation and mitigation and for generating reliable funding for the purpose". By not applying the PPP while opting for the UNFCCC's Article 3.1

on common but differentiated responsibility based on respective capabilities (customised in the Paris Agreement Article 2), the NDP takes the position that it is not the main GHG emitters' primary responsibility to advance and fund climate action. This is a foundational weakness of the NDP, which as demonstrated further below, partly accounts for the country's climate governance deficiencies and failures.

While the NDP is the current overarching macro NDP, South Africa has adopted a coterie of climate policies, strategies and plans for mitigation and adaptation since its ratification of the UNFCCC in 1997. Within the international climate governance regime, the country's mitigation and adaptation obligations, as per the requirements of the Paris Agreement, are articulated in its NDC. On the mitigation front, the country's successive NDCs from 2015 reiterate the need to reduce emissions with absolute targets of "between 398 to 614 Mt CO_2 reduction over the period 2025–2030 and an additional 212 to 428 Mt CO_2 reduction by 2050" (Petrie *et al.* 2018: 27). These GHG reduction targets are, however, low in view of the country's relatively high per capita CO_2 emissions per annum. In fact, South Africa's NDCs have been criticised for not being ambitious and aggressive enough in their mitigation commitments and targets. "The mitigation component of South Africa's NDC is considered inconsistent with the Paris Agreement under any metrics of ambition (accounting for capacity to pay, historical responsibility or population)" (Robiou du Pont *et al.* 2017 cited in Petrie *et al.* 2018: 28). Yet, despite these low-emission reduction targets, it is unlikely they will be met (previously, the country failed to reduce GHGs by 34% by the end of 2020) due to the country's languid approach in implementing mitigation activities. More recently, Ziervogel *et al.* (2022: 1) have argued that the country's mitigation targets "remain inadequate and the implementation of these promises even slower". This 'snail-pace' of implementing inadequate targets demonstrates lack of climate action ambition and overall failure to aggressively implement mitigation strategies.

This failure becomes starker when one assesses the country's flagship climate action strategies, namely, the NCCRS, NCCRP and NCCAS. While the NCCRS (GovSA 2004a) acknowledges the North's role in mitigation action in developing countries, this responsibility is framed in policy and programming nomenclature that is not definitive, ahistorical and without climate justice lens. Its argument that "the developed nations of the world, with their immense capital reserves, need to be encouraged to develop appropriate technologies to mitigate global climate change" which "South Africa, as an integral part of the developing world, is always willing to accept" (GovSA 2004a: vii–viii) demonstrates its inability to properly frame national climate governance within the context of climate justice. It does not take full cognisance of the historical anthropogenic development of the climate emergency that primarily happened due to the Industrial Revolution, state-led industrialisation, capital accumulation and unhinged consumerism in the North. It also fails to situate its strategic direction on the reality that the North has a historical responsibility to shoulder the burden of supporting just climate actions and just transition to climate-resilient development in South Africa. This historical responsibility view emerges from a Global South consensus that the role of the North is not just to develop climate-smart mitigation technologies for export (for profit) but rather to primarily reparat through practical climate action support that is just, fair and equitable. This is the North's primary responsibility because their developed economies were built on uninhibited environmental destruction, high GHG-emitting coal, oil and fossil fuels exploitation at the expense of the South.

Alongside the NCCRS is the NCCRP. The NCCRP outlines the multisector framework for mitigation and adaptation (GovSA 2011a). It derives its vision of a low-carbon and climate-resilient future from its predecessors, namely, the NCCRS and the LTMS. It is instructive that this policy has elements of climate justice thinking. This is embodied in its recognition of the fact that climate action must meet the special needs and circumstances "of localities and people that are particularly vulnerable to the adverse effects of climate change" with a view to uplifting them to "ensure human dignity" and equity (GovSA 2011a: 12). Some of the policy's ideas are thus embedded in climate justice thinking in as far as they aspire for actions that are fair, equitable and address the vulnerabilities of the poor and marginalised. However, the extent to which this climate justice leaning macro-policy has or is contributing to fair, equitable and effective responses that concurrently build local agency, resilience, drive poverty reduction and diversified livelihood activities remains contested. There are numerous reasons for this contestation. Firstly, the NCCRP is full of contradictions, internal dissonance and inconsistencies. The first contradiction is that its 'common but differentiated responsibilities and respective capabilities' principle directly resonates with neoliberal globalisation thinking characteristic of UNFCCC and Paris Agreement language and reasoning. This neoliberal position is not in harmony with some of its climate justice elements mentioned above. Secondly, it contradicts itself through attempting to jointly employ the 'common but differentiated responsibilities and respective capabilities' approach and the PPP. The two, PPP and 'common but differentiated responsibilities', are mutually exclusive in climate justice discourse. This is because PPP is unapologetic about the fact that the North must pay for 'loss and damages' caused by the climate emergency while the second principle is ambivalent about this fact. Thirdly, on the mitigation front, the NCCRP's approach is informed by the geopolitical desire of the country to be seen as a responsible global citizen. This geostrategic interest accounts for the country's rather extravagant proclamation that GHG emissions will decline in absolute terms from 2036 onwards. As discussed above, there is no credible action plan or evidence to suggest that these emission reduction targets will be progressively achieved. There is also no nationally comprehensive, coordinated plan of how provinces, including the Eastern Cape where Buffalo City is located, can collectively contribute to GHG emission reductions in a manner and process that can be monitored and evaluated for compliance.

Another impediment to full implementation of climate actions is lack of a national climate policy implementation framework to guide local-level activities. Petrie *et al.* (2018: 26) note that within South Africa's local government system, "mandates in regards to climate change are unclear and widely interpreted" with consequences on the choice, practicality and impact of interventions. For example, while the NCCRP (2011a) calls for coordination of climate actions across the three tiers of the government, with citizens exercising their formal and informal power in all processes through active participation, it does not clarify the specific climate action roles and responsibilities of local governments within this inter-governmental matrix. As a result, municipalities are proceeding guided by their own interpretation of national climate policies and strategies as opposed to reliance on a synergistic national framework. Had a national implementation framework been in existence, municipalities' programme of action would be guided with clear, quantifiable targets and outcomes. Relatedly, local government legislation (e.g., the Municipal Systems Act and Municipal Structures Act) also does not specify local governments' climate action mandate. Both Acts make reference to promotion of a 'safe and healthy environment' which municipalities broadly interpret to encom-pass and guide their climate programming. This means the absence of a policy and legal implementation framework that gives municipalities guidance to design and implement specific climate mitigation and adaptation plans partly explains their piecemeal approach to the climate emergency.

Further, an assessment of different tiers of governments' mitigation actions has crystallised the consensus that the country's "current climate policies and actions are insufficient" to drive it towards "the 1.5°C temperature limit" (Climate Action Tracker 2022: 1) due to its continued dependence on coal. The negligible progress in reducing reliance on coal for energy partially emanates from climate policy and strategic plans dissonance. While the NDP and NCCRP argue for a gradual shift to renewable energy sources, the IRP of 2019 directly contradicts this policy position. It argues that "coal will continue to play a significant role in electricity generation in South Africa in the foreseeable future as it is the largest base of the installed generation capacity and it makes up the largest share of energy generated" (GovSA 2019b: 12). It adds that due to "the abundance of coal resources, new investments will need to be made in more efficient coal technologies (HELE technology, includ-ing supercritical and ultra-supercritical power plants with CCUS) to comply with climate and environmental requirements" (GovSA 2019b: 12). This stance on coal by the GovSA demonstrates that there are no reasonable prospects for the country to achieve its GHG emission reduction targets by 2030. The country's economy will remain coal dominated, thus continuing high GHG emissions while simulta-neously serving the profit and accumulation interests of mining conglomerates at the expense of climate actions geared towards achieving sustainable and equitable resilience for the poor and marginalised most affected by the climate emergency.

There is also lingering uncertainty about whether, and the extent to which, sector-specific mitigation strategies such as the GTS will be implemented fully. The GTS is founded on evidence that recognises that direct and indirect "emissions from the transport sector in South Africa account for 10.8% of the country's total GHG emissions" (GovSA 2018: 5) with significant impact on the climate. To mitigate and adapt to this, its action plans revolve around promotion of "energy efficiency and emission control measures", thus transitioning the sector to a low-carbon future, "promoting behavioral changes towards sustainable mobility" and "facilitating the sector's just transition to a climate resilient transport system and infrastructure" (GovSA 2018: 6). However, while these transition plans make strategic trajectory sense, they are not (yet) backed by significant finance to drive the transition. A humongous sector like transport, which is central to economic activities, requires huge investment to enable a balanced transition to green transport and green mobility. Without this investment, the GTS thus remains another good strategy not yet fully implemented in the climate governance space. Similarly, other strategic plans such as Green Growth which permeate not just the NDCs but crosscut other sector-specific strategies illustrate mitigation action failures. Despite Green Growth commitments and an enabling policy environment, a number of Green Growth-linked mitigation targets remain unfulfilled. For example, the Green Economy Accord of 2011, a social partnership of government, business, labour and community representatives, pledged "to create 300 000 new [green] jobs by 2020" (GovSA 2011b: 1) but failed to achieve this target. This is because there was no meaningful public and private investment in renewable energy generation, biofuels production and eco-tourism sectors earmarked as drivers of these new green jobs.

Outside government policy spaces, there are financial power dynamics that also hinder full implementation of climate policy and strategies, especially those that relate to a just transition. The profit motive, particularly of some big commercial banks, is partly derailing this transition and decarbonisation agenda because they have a vested interest in maintenance of the coal and fossil fuels industry status quo since they are generating enormous profits through loans and investments. For example, the Standard Bank Group of South Africa argues that it will continue financing "exploration, extraction and processing of thermal coal" projects, specifically those of "existing coal fired power generation utilities" (Standard Bank 2020: 3). This coal financing stance, regardless of its perpetuation of high GHG emissions, is reiterated by the Bank's Chief Executive Officer of Corporate and Investment Banking, Kenny Fihla, who emphasises that "energy generation in South Africa and Sub-Saharan Africa in particular, is still more than 80% driven by coal fired power stations, so clearly, in that scenario, we will continue to support the generation of electricity" (Mashego 2020: 1). Evidently, Standard Bank's continued financing of the coal sector is an investment decision driven by neoliberal financial thinking and its associated pursuit of profits for shareholders regardless of its climate system damage and perpetuation of climate injustice. It also 'flies in the face' of the government's climate agenda to drive the country towards a net-zero carbon economy. This push back by big capital, and condescension to decarbonisation, is mirrored in the ABSA banking group as well. Although ABSA supports green energy projects through a certified green loan provided by the International Finance Corporation, finances renewable energy technology, solar, wind and hydropower generation and has a coal financing standard with capital caps (ABSA 2022), these are of limited breadth and impact when viewed within the context of the national climate emergency. Further, even though ABSA states that it "will not fund new coal-fired electricity generation" (ABSA 2020: 1), the fact that it remains a major financier of ongoing coal and fossil fuel businesses and is open to future investment in this sector "under extenuating circumstances that will be governed under strict guidelines" (ABSA 2020: 1) demonstrates its half-hearted tactic to climate actions that are pathway to a just transition. Regardless of its few sustainability and climate financing actions, it remains indisputable that

ABSA's contribution to a just transition and national net-zero carbon emission targets remains limited. This is due to its continued active participation, in pursuit of profit, in the coal and fossil fuels sector.

The foregoing means South African banks are part of the problem as they are lagging behind in mitigating climate risk and supporting adaptation actions for a transition to a low-carbon sustainable economy. This lethargy emanates from a disconnect between their well-written glossy climate policies and actual actions. On the one hand, some of them proclaim discontinuation of coal and fossil fuel financing, while, on the other hand, they argue that extenuating unusual circumstances compel them to extend new financing. It gets worse; those coal and fossil fuel industries with long-term loans and investments are excluded from this discontinuation policy. Clearly, the bottom line here is that Banks are not driven entirely by a desire to ensure that their loans and investments are structured towards emission reductions; instead, their overriding motive is profiteering for shareholders, hence their continuation on a path with minimal financial risk. This commercial behaviour by Banks is aptly captured by Rennkamp's (2019: 755) observation that some of "the most contested climate policies are those that create distributional conflicts where powerful, non-poor actors will potentially experience real losses to their fossil fuel based operations". The lending behaviour of Banks, despite their few sustainability and climate actions, should therefore be read within this financial and economic power matrix that has little regard for advancing realisation of climate justice and a just transition.

From Policy to Practice: Climate Governance, Inactions and Injustice in Buffalo City

Buffalo City is the second-largest city in the Eastern Cape Province. Therefore, this exploration of its climate governance agenda and actions should be understood within the context of the provincial governance matrix. This means the Eastern Cape Climate Change Response Strategy (ECCCRS) (Eastern Cape Department of Economic Development and Environmental Affairs [ECDEDEA] 2011) is one of the pillars on which Buffalo City's climate action agenda hinges on. The ECCCRS' strategic position is that mitigation and adaptation actions within the province can only be feasibly implemented through departmentally integrated response programmes "where multiple sectors and departments contribute to a common climate change issue to ensure effective adaptation responses" (ECDEDEA 2011: 5). This inter-departmental coordination and liaison means the analysis of Buffalo City's mitigation, adaptation and resilience programmes as vehicles for climate justice and just transition is thus being done within this provincial context that is also subsumed by national climate policies and actions. Further, prevailing provincial polity and climate politics should be kept in mind as they partly explain the context and scope of climate governance at the city level. Polity, hereunder, refers "to the broader institutions, and sets of formal and informal rules that structure the way political processes occur [while] politics refers to the political actors and their power resources and relationships" (Rennkamp 2019: 757). Hence, provincial polity on climate action and local Buffalo City politics mediate climate governance strategic priorities, actions and outcomes. In other words, climate governance outcomes at the city level are determined by the interactions of polity and politics. This is re-iterated by Rennkamp's (2019: 757–758) argument that "the broader institutional regime (the polity) as well as power relations, political negotiation processes and shaping of coalitions (politics) explain policy outcomes" in the climate action space.

The polity and policy matrix at the city level thus manifest in Buffalo City's climate actions and omissions, both of which have a bearing on climate justice. On paper, the city's Climate Change Strategy (adopted in 2015) is organically embed-ded in its current Integrated Development Plan (IDP) (2021–2026) strategic objec-tive which aims to "take urgent action to combat climate change and its impacts" (Buffalo City Metropolitan Municipality [BCMM] 2021: 33) as a pathway towards a green city. While this city strategic objective is aligned to the Provincial Develop-ment Plan's (PDP) vision of "a growing, inclusive and equitable economy" (PDP cited in BCMM 2021: 35) and the NDP's idea of pursuing "environmental sustain-ability and resilience through an equitable transition to a low-carbon economy" (GovSA 2012 cited in BCMM 2021: 27), there is no evidence of tangible integrated city-level climate actions towards realisation of resilience and an equitable low-carbon economy. This is due to the fact that its integrated development planning and climate strategy statements of intent are not accompanied by a realistic climate governance framework and strategies of climate action. In other words, the city currently does not have a practical climate governance framework to strategically guide climate actions with specific milestones and targets. This lack of a functional framework partly explains why it always struggles (and fails) to respond to extreme climate events such as recurrent floods. For example, the city's weak, perhaps dys-functional, adaptation and resilience-building systems were exposed in January 2022 when it dismally failed to marshal its early warning systems, equipment, resources and human capital to effectively respond to climate-induced flooding in poor, working-class communities. Consequently, heavy flooding in Mdantsane and Duncan Village townships resulted in internal displacement of poor people, de-struction of their homes, household property and public infrastructure such as roads and bridges (Brand and Solomons 2022; Sgqolana 2022). This climate disaster management response failure of the city is hardly surprising. By its own admission, the city's "disaster management is not practiced in an integrated and coordinated, multi-sectoral, risk focused manner" because there is "insufficient disaster man-agement capacity and budgeting" as well as "lack of understanding regarding fund-ing for disaster management activities" (BCMM 2021: 172). This disorganisation at the official level fully explains why climate shocks such as the recent floods were not anticipated by the city's disaster management teams, hence dysfunctionality in the *ex post* adaptation response efforts. The poor and working class who live in the townships of Mdantsane and Duncan Village were disproportionately affected, further demonstrating how they continue to suffer climate injustice. This injustice also epitomises the inequity of climate impacts within a city because while there was heavy rainfall throughout the city, more affluent neighbourhoods with better drainage systems and landscaping were not flooded, while poor townships were the epicentre of flooding destruction and displacement.

In addition, while the city has mainstreamed climate mitigation and adaptation in its planning through horizontal alignment of its long-term Metro Growth and Development Strategy, current IDP and Climate Change Strategy, this is, however, undone by poor climate action programming and lack of climate finance. For example, while the city acknowledges that disaster management law and procedures require it to budget for climate change *ex ante* mitigation and *ex post* adaptation, especially for vulnerable groups (BCMM 2021), this is hardly ever done to satisfaction as very limited financial resources are budgeted for that purpose. This means, despite climate mainstreaming at the planning level, strategic climate action implementation is poor, and highly fragmented, due to lack of an implementation budget. One can add that mainstreaming climate action on paper without committing financial and technical resources for practical implementation defeats the primary objectives of the city's Climate Change Strategy and the IDP. The two, planning and budgeting, are symbiotic twins in any climate action endeavour that seeks to address inequitable climate impacts and build resilience of the poor and vulnerable.

Further, the city's lackadaisical climate financing is visible in peri-urban farming communities where it is failing to enhance local-level farmer-initiated adaptation and resilience-building efforts. For example, while the city partially funded and installed irrigation systems at Mzintshana, Mzantsi and Pirie Mission local communities and supported some "emerging farmers with poultry and piggery infrastructure [in Wards 37 and 40] to the value of R1 200 000" (BCMM 2021: 159) in 2020–2021, evidence gathered during a transact walk in these communities demonstrated the weaknesses of a piecemeal approach in building local resilience. While these farmers are deserving beneficiaries of this climate finance largesse, many equally eligible are not recipients of this assistance. This means the climate finance and technical support provided was done in a disjointed manner as there was no evidence of universal provision to these or other communities prior and after 2021. It appears the city is not actively implementing this support programme as part of a long-term peri-urban farming adaptation and resilience-building initiative. Rather it is continuing on a 'business as usual' approach without due regard to inequitable climate impacts in the peri-urban subsistence farming community. This fragmented approach means its climate financing and technical support is not sustainable, hence without a realistic chance of addressing farmers' underlying vulnerability and inability to respond to inequitable climate impacts. Not only is this an indictment of the city's climate governance approach but also a demonstration of how it is failing to systematically implement strategic adaptation interventions with potential to drive these communities towards climate justice.

On the other hand, some may argue that there is a caveat to these failures by the city. It is worth remembering that there are national and provincial polity and policy dynamics that mediate the ability or inability of the city to deliver appropriate and just climate actions. The intersection of power, polity and institutional authority embedded in South Africa's constitutionally separate yet interrelated spheres of national, provincial and local governments determines the extent to which a local authority like Buffalo City can implement human development programmes, including climate action programming. This hierarchical power and authority, cascading from national down to local, means local governments such as Buffalo City have the least resources to determine climate financing and proofing programming priorities even though they are at the forefront of local human development and climate action. The political power and administrative authority delegated to national- and provincial-level office bureaucrats means they yield more power and resources to determine and shape climate financing and proofing programming priorities. Such a system of multilevel climate governance suggests that we should perhaps be more circumspect when analysing the extent to which Buffalo City can deliver effective and sustainable climate actions given that it does not possess or control infinite resources to drive mitigation and adaptation activities in local communities.

There is also a need to understand the extent to which the relationship between political power and climate leadership influences climate action within the city. Although the concept of climate leadership is contested, the working narrative herein is that an actor or institution qualifies as a climate leader if their climate actions are "more ambitious than others in the pursuit of the common good", specifically in relation to the multilaterally agreed Paris Agreement's long-term mitigation, adaptation and resilience targets (Oberthür and Dupont 2021: 1097). Therefore, actors or institutions are climate leaders if they pursue and implement ambitious and exemplary climate actions (compared to other actors or institutions) towards globally agreed objectives. In the context of Buffalo City, such exemplary actions should be towards achieving climate justice for the poor and vulnerable. The question then is, to what extent does Buffalo City as an institution deserve this positive normative connotation of climate leadership? To fully answer this question, one must step back and unpack the relationship between local political power dynamics, political priorities and climate action within the city. The fact is local political priorities with their accompanying political capital primarily shape the direction of the city's integrated development planning and programming. Hence, institutional decision-making within the city to actively implement climate action (or not) is directly influenced by these political calculations. In cases where political opportunities and political mileage increase through the construction of low-cost housing and provision of basic services for the poor and vulnerable, the city's political and local government leaders channel resources and time in that direction. Consequently, mitigation, adaptation and resilience building are pushed to a secondary agenda, thus demonstrating lack of exemplary climate leadership in the city. This means that political considerations in the city's corridors of power are limiting its ability to be ambitious in pursuit of climate actions for the common good, especially climate justice for the poor and vulnerable. Without exemplary climate leadership, climate actions within the city will therefore remain at the periphery of integrated development planning and programming, thus further diminishing the chances of achieving climate justice.

Beyond seeing political opportunism as one of the explanatory factors for lack of climate leadership in the city, perhaps it is important to reflect on another perspective of why political and local government leaders are prioritising basic service delivery for the poor and vulnerable instead of climate action. It is well documented that Buffalo City suffers from limited local economic development, chronic poverty, high unemployment, lack of adequate public infrastructure and basic services. These are visible human development challenges confronting the poor and working class in their daily lives. Political and local government leaders therefore prioritise resolution of these existential challenges in response to citizens' demands and because perpetual failure to address these carries a political risk for the local government of the day. As a result, climate action is either postponed, ignored or simply put on the back burner. This relegation of climate actions to secondary consideration is hardly surprising given that the local governments' political survival at the polls is judged on the delivery of basic services rather than climate leadership. On the other hand, prioritisation of service delivery only also demonstrates how local government leaders and politicians are failing to create an ecosystem that links their integrated development mandate, political survival scheming and climate leadership. Yet this is possible through, for example, pursuing mitigation and adaptation programmes that drive 'green' economic activities, create local employment opportunities and advance the local development agenda. Such co-benefits are possible but can only be achieved under a climate conscious local government and political leadership with the capacity and drive for steering climate policy and action towards climate justice.

These glaring failures of the city to consistently and effectively implement sustainable climate adaptation and resilience actions lay bare the limitations of a monocentric regime in climate governance and action at the city level. This failed state-centric, top-down approach (that extends beyond South Africa) has encouraged the growing influence of non-state actors in climate action. At a global level, there is increasing involvement of "non-state actors in the governance of climate change" (Okereke *et al.* 2009: 58), giving rise to a polycentric climate governance system. The advent of this "polycentric governance offers new opportunities for climate action", founded on its underlying ontology of bottom-up, multilevelled and diverse governing that has "a better fit with local priorities" (Jordan *et al.* 2015: 977–978). In line with this polycentric governance, non-state actors, primarily non-governmental organisations (NGOs) and community-based organisations (CBOs), now span the breadth of the Global South implementing project and programme-based climate mitigation, adaptation and resilience interventions. They are now also a central element of the climate governance architecture in South Africa, including in Buffalo City.

Equipped with a bottom-up and a more people-centred governance framework, NGOs such as Gift of the Givers are now prominent in climate disaster relief and recovery in Buffalo City. For example, Gift of the Givers was central in the 2022 *ex post* adaptation to flooding in the townships of Mdantsane and Duncan Village. These floods swept away homes and household belongings, destroyed public health facilities and schools infrastructure leaving six people dead "including a police officer who was conducting rescue and recovery operations at the time" (Francke 2022: 1). In response, Gift of the Givers arranged temporary housing for all affected residents and provided sleeping paraphernalia and meals. Although there are many untested assumptions (and contested conclusions) about the performance, effectiveness and sustainability of NGO work in climate action (Jordan *et al.* 2015: 977), it is a truism that they are, in principle, the vanguards and pro-poor actors. They serve, as Gift of the Givers did, the poor and vulnerable who are underserved, ignored or abandoned by government entities when climate crisis-induced disasters are at their most ferocious. Their increasingly noticeable interventions also point to the gradual decline of a failing monocentric top-down approach as a more pluralistic polycentric form of climate governance gains traction. While exercising cautious optimism, the empirical reality provided by the intervention of Gift of the Givers in Buffalo City validates the argument that a polycentric pattern is perhaps the best avenue going forward for complementary climate action between the city and non-state actors. The advantages of such a polycentric approach needs to be researched in other South African cities in order to further understand the new opportunities provided by NGOs that are, in principle, free of bureaucratic bottlenecks, are relatively well managed and resourced thus in prime programming position to fill climate action gaps left behind by city governments. It will also be worthwhile to explore the power dynamics between non-state and state actors in climate governance as well as the geostrategic tensions arising out of the growing influence of these Global North donor-funded non-state actors and agencies of the state in urban South Africa.

Conclusion

At the centre of Buffalo City's failed climate governance system is the fact that its climate conscious IDPs and climate action strategies remain 'on paper' without noteworthy practical measures or steps being taken towards full implementation. This failure emanates from the intersection of polity and politics that have a direct bearing on how city climate action priorities are decided. The three-tier multilevel climate governance structure means the city is at the bottom of the political power ladder, decision-making authority and access to resources. Yet the city is expected to drive effective and sustainable mitigation and adaptation activities when they have minimal, and perhaps, pseudo power and the least climate finance muscle. Relatedly, these power dynamics extend to the internal processes within the city itself. Where polity and political dynamics favour pursuance and delivery of traditional basic services and infrastructure in order to retain power at the city hall, then these considerations supersede the need for climate actions in response to the climate emergency. Although this might appear, at face value, to be a noble decision taken in the context of competing interests between local development needs and climate action, it is in fact symptomatic of the city's failure to build an integrated development ecosystem that meets local human development needs while simultaneously driving a mitigation, adaptation and resilience agenda towards climate justice.

Further, despite mainstreaming climate action in its local development plans and strategies, the city is failing to adequately budget and allocate climate finance. Consequently, climate programming remains fragmented as the city is failing to effectively implement *ex ante* mitigation and *ex post* adaptation, especially in poor and vulnerable communities, thus deepening climate injustice. The human cost of this has been loss of lives in poor communities, destruction of their homes, damage to their movable household property and public infrastructure due to flooding. Yet all these ferocious impacts could have been avoided, or at the very least minimised, if there was climate leadership in the city. Lack of climate leadership, i.e., political and bureaucratic actors with the ability to spearhead exemplary climate actions, is holding back the city's climate agenda as there is preoccupation with political capital and power retention rather than pursuing climate-resilient action for the benefit of the poor and vulnerable. Therefore, without exemplary climate leadership, climate actions within the city will remain at the periphery, thus diminishing the chances of achieving climate justice. Lastly, the climate governance failures of the city are also symptomatic of the broader weaknesses of a monocentric, top-down approach when attempting to respond to a global common challenge like the climate emergency. The limitations of the idea that the state, with all its political and bureaucratic imperfections, can unilaterally implement mitigation, adaptation and resilience actions at a city level has been laid bare in Buffalo City. It is impractical for the city to go it alone hence the advent and growing programming popularity of a polycentric climate governance system. This system that is built on complementing the strengths of state and non-state actors in climate action offers new opportunities for the future. With well resourced, managed, pro-poor NGOs as its driving force, this bottom-up pluralistic system is gaining traction as a forward-looking climate governance approach that offers a pathway towards climate justice for the poor and vulnerable in cities.

References

ABSA (2022) '2021 Task Force on Climate-Related Financial Disclosures Report', Johannesburg (available at https://www.absa.africa/content/dam/africa/absaafrica/pdf/sens/2021/2021-Absa-Group-Limited-TCFD.pdf).

ABSA (2020) 'Absa Group Publishes Sustainability Policy and Standard on Coal Financing', Johannesburg (available at https://www.absa.africa/media-centre/media-statements/2020/summary-coal-financing-standard/#:~:text=As%20a%20result%2C%20Absa%20will,be%20governed%20under%20strict%20guidelines).

Andonova, L.B., Betsill, M.M. and Bulkeley, H. (2009) 'Transnational Climate Governance', *Global Environmental Politics,* Vol 9 (2), 52–73, https://doi.org/10.1162/glep.2009.9.2.52.

Atwoli, L., Baqui, A.H., Benfield, T., Bosurgi, R., Godlee, F., Hancocks, S., Horton, R., Laybourn-Langton, L., Monteiro, C.A., Norman, I., Patrick, K., Praities, N., Rikkert, M.G.M.O., Rubin, E.J., Sahni, P., Smith, R., Talley, N., Turale, S. and Vázquez, D. (2021)

'Call for Emergency Action to Limit Global Temperature Increases, Restore Biodiversity, and Protect Health: Wealthy Nations Must Do Much More, Much Faster', *Nutrition Reviews*, Vol 79 (11), 1183–1185, https://doi.org/10.1093/nutrit/nuab067.

Brand, A. and Solomons, L. (2022) 'Buffalo City Activates Disaster Management Teams amid Heavy Floods in East London', *News24*, 8 January (available at https://www.news24.com/news24/southafrica/news/buffalo-city-activates-disaster-management-teams-amid-heavy-floods-in-east-london-20220108).

Buffalo City Metropolitan Municipality (2021) *Integrated Development Plan 2021/26* (available at https://www.buffalocity.gov.za/idp.php).

Climate Action Tracker (2022) 'South Africa: Policies and Action' (available at https://climateactiontracker.org/countries/south-africa/policies-action/).

Coen, D., Kreienkamp, J. and Pegram, T. (2020) *Global Climate Governance*, Cambridge: Cambridge University Press.

Eastern Cape Department of Economic Development and Environmental Affairs (2011) *Eastern Cape Climate Change Response Strategy*, Bhisho: Eastern Cape Department of Economic Development and Environmental Affairs.

Francke, R.L. (2022) 'Gift of the Givers Assist Flood Victims in East London', *Independent Online*, 10 January (available at https://www.iol.co.za/news/south-africa/eastern-cape/gift-of-the-givers-assist-flood-victims-in-east-london-9ce9a3ca-dec9-45b7-be86-38adc31d9aa6).

González, S.C. and Numer, E. (2020) 'What Is Climate Governance?' United Nations Children's Educational Fund, Latin America and Caribbean Regional Office, Panama (available at https://www.unicef.org/lac/media/19651/file/what-is-climate-governance.pdf).

Government of South Africa (2019a) *Carbon Tax Act No. 15 of 2019* (available at https://www.gov.za/documents/carbon-tax-act-15-2019-english-afrikaans-23-may-2019-0000).

Government of South Africa (2019b) *Integrated Resource Plan 2019* (available at http://www.energy.gov.za/files/irp_frame.html).

Government of South Africa (2018) *Green Transport Strategy for South Africa: 2018–2050* (available at https://www.transport.gov.za/documents/11623/89294/Green_Transport_Strategy_2018_2050_onlineversion.pdf/71e19f1d-259e-4c55-9b27-30db418f105a).

Government of South Africa (2014) *Medium-Term Strategic Framework* (available at https://www.gov.za/sites/default/files/gcis_document/201409/mtsf2014-2019.pdf).

Government of South Africa (2012) *National Development Plan 2030: Our Future-Make It Work* (available at https://www.gov.za/sites/default/files/gcis_document/201409/ndp-2030-our-future-make-it-workr.pdf).

Government of South Africa (2011a) *National Climate Change Response White Paper* (available at https://www.gov.za/sites/default/files/gcis_document/201409/nationalclimatechangeresponsewhitepaper0.pdf).

Government of South Africa (2011b) *South Africa's Green Economy Accord* (available at https://www.gov.za/south-africas-green-economy-accord).

Government of South Africa (2008a) *National Energy Act 34 of 2008* (available at https://www.gov.za/documents/national-energy-act#:~:text=to%20provide%20measures%20for%20the,for%20all%20matters%20connected%20therewith).

Government of South Africa (2008b) *National Environmental Management: Integrated Coastal Management Act 24 of 2008* (available at https://www.gov.za/documents/national-environmental-management-integrated-coastal-management-act#:~:text=to%20prohibit%20incineration%20at%20sea,relation%20to%20coastal%20matters%3B%20and).

Government of South Africa (2004a) *A National Climate Change Response Strategy for South Africa*, Pretoria: Department Of Environmental Affairs and Tourism.

Government of South Africa (2004b) *National Environmental Management: Biodiversity Act 10 of 2004* (available at https://www.gov.za/documents/national-environmental-management-biodiversity-act-0).

Government of South Africa (2002) *Disaster Management Act 57 of 2002* (available at https://www.gov.za/documents/disaster-management-act).

Government of South Africa (1996) *The Constitution of the Republic of South Africa* (available at https://www.gov.za/documents/constitution/constitution-republic-south-africa-1996-1).

Intergovernmental Panel on Climate Change (2022) 'Climate Change 2022: Mitigation of Climate Change', IPCC Sixth Assessment Report (available at https://www.ipcc.ch/report/ar6/wg3/).

Jordan, A.J., Huitema, D., Hildén, M., Van Asselt, H., Rayner, T.J., Schoenefeld, J.J., Tosun, J., Forster, J. and Boasson, E.L. (2015) 'Emergence of Polycentric Climate Governance and Its Future Prospects', *Nature Climate Change*, Vol 5, 977–982, https://doi:10.1038/NCLIMATE2725.

Khan, M.R. (2015) 'Polluter-Pays-Principle: The Cardinal Instrument for Addressing Climate Change', *Laws*, Vol 4, 638–653, https://doi:10.3390/laws4030638.

Lindsey, R. and Dahlman, L. (2022) 'Climate Change: Global Temperature', Science and Information for a Climate Smart Nation (available at https://www.climate.gov/news-features/understanding-climate/climate-change-global-temperature).

Mashego, P. (2020) 'Standard Bank Commits to Funding Coal Mining in South Africa and Region', FIN24, 10 December (available at https://www.news24.com/fin24/companies/standard-bank-commits-to-funding-coal-mining-in-south-africa-and-region-20201210).

Oberthür, S. and Dupont, C. (2021) 'The European Union's International Climate Leadership: Towards a Grand Climate Strategy?' *Journal of European Public Policy*, Vol 28 (7), 1095–1114, https://doi.org/10.1080/13501763.2021.1918218.

Okereke, C., Bulkeley, H. and Schroeder, H. (2009) 'Conceptualizing Climate Governance Beyond the International Regime', *Global Environmental Politics*, Vol 9 (1), 58–78. https://doi.org/10.1162/glep.2009.9.1.58.

Petrie, B., Wolpe, P., Reddy, Y., Adriázola, P., Gerhard, M., Landesman, T., Strauch, L. and Marie, A. (2018) *Multi-Level Climate Governance in South Africa. Catalysing Finance for Local Climate Action*, Berlin and Cape Town: OneWorld, Sustainable Energy Africa (available at https://www.adelphi.de/en/publication/multi-level-climate-governance-south-africa).

Rennkamp, B. (2019) 'Power, Coalitions and Institutional Change in South African Climate Policy', *Climate Policy*, Vol 19 (6), 756–770, https://doi:10.1080/14693062.2019.1591936.

Sgqolana, T. (2022) 'Eastern Cape Officials Still Counting Losses after Floods Kill 14 and Displace Hundreds', *Daily Maverick*, 12 January (available at https://www.dailymaverick.co.za/article/2022-01-12-eastern-cape-officials-still-counting-losses-after-floods-kill-14-and-displace-hundreds/).

Standard Bank (2020) 'Fossil Fuels Financing Policy' (available at https://www.standardbank.com/static_file/StandardBankGroup/filedownloads/PolicySummary_FossilFuelsFinancing.pdf).

United Nations (2015) *The Paris Agreement* (available at https://unfccc.int/process-and-meetings/the-paris-agreement).

Ziervogel, G., Lennard, C., Midgley, G., New, M., Simpson, N.P., Trisos, C.H. and Zvobgo, L. (2022) 'Climate Change in South Africa: Risks and Opportunities for Climate-Resilient Development in the IPCC Sixth Assessment WGII Report', *South African Journal of Science*, Vol 118 (9/10), https://doi.org/10.17159/sajs.2022/14492.

9   Precarious Adaptation to Climate Impacts

Farmer Agency, Choices and Constraints in the Eastern Cape, South Africa

Philani Moyo

Introduction

Climate change impacts and their existential threat to humanity in rural and urban South Africa are well documented. They include erratic rainfall patterns, high temperatures (Tshiala et al. 2011; Zikhali et al. 2020), recurrent droughts (Engelbrecht and Monteiro 2021), severe floods, heat waves and wildfires (Schulze 2016). The severity and frequency of these impacts is not uniform across all geographic regions of South Africa. There are spatial variations at provincial, district and local community levels. For example, parts of KwaZulu Natal have higher flood risk and incidents (Department of Cooperative Governance and Traditional Affairs 2019; Mkhize 2022) compared to recurrent high temperatures and droughts in Limpopo (Nyoni et al. 2021), parts of the Northern Cape and Eastern Cape. Similarly, different geographic regions of the Western Cape Province regularly experience flooding (Dube et al. 2021), while others are prone to droughts (Otto et al. 2018; Burls et al. 2019). These geospatial variations mean climate impacts vary inter and intra provincially as well as locally. Despite spatial variabilities, wherever they occur, the consequences of climate change are disproportionately profound for rural inhabitants since they are negatively affecting agriculture production, food and nutrition security (Zikhali et al. 2020), water security (Engelbrecht and Monteiro 2021), human health (Chersich et al. 2018), infrastructure, human settlements, ecosystems and biodiversity.

The differential and uneven effects across households are the product of vulnerability that is "a function of exposure, sensitivity and adaptive capacity" (Moser et al. 2010, 2). Primarily, two forms of vulnerability, external and internal, mediate households' capacity to adapt and build resilience to climate impacts. "The first is an external dimension that comprises the risks, shocks, and stresses to which people are subject; and second, an internal dimension that encompasses their capacity and associated means to withstand, or adjust, to damaging losses" (Moser et al. 2010, 6). Within, and between, households, varying resources, asset portfolios, infrastructure, expertise, access to climate information and services influence a households' capacity to adapt to adverse climate impacts. Other socio-economic

DOI: 10.4324/9781003397120-11

drivers of internal vulnerability "include the gender, age, health, social status, and ethnicity of individuals" (Noble et al. 2014, 836). These differing vulnerability levels mean some of the households most susceptible to climate risks and shocks are those with the least adaptation assets, skills and information. In rural South Africa, the majority of smallholder farmers belong to this vulnerable cohort. These smallholder farmers on communal two hectares or less of land (Livingston et al. 2011 in Zikhali et al. 2020) are largely subsistence oriented due to the productivity limits of their small landholdings, limited access to requisite farming inputs, resources, agricultural technical skills and information. Within this context, it is important to unravel and understand their agency, active adaptation to the raging climate crisis and the obstacles they encounter in the process. This chapter, however, moves beyond simply analysing farmers' actions in response to climate impacts to an analysis of how underlying vulnerability drivers mediate adaptive capacity and resilience building.

With reference to purposively sampled smallholder farmers in Mqanduli communal area in the Eastern Cape Province of South Africa, the primary aim of this article is to assess adaptation obstacles they endure and their response actions to these challenges as they seek to build resilience. This chapter answers the following primary research questions: (1) What are the underlying drivers of adaptation constraints faced by smallholder farmers? (2) To what extent are assets (different forms of capital) ameliorating farmers' adaptation challenges as they seek to build resilience? (3) To what extent is the provincial government's material and technical support enabling smallholder farmers to effectively adapt and build resilience? Answers to these and secondary questions provide an entry point for understanding whether current smallholder farming practices and actions are enabling them to effectively adapt. Further, an assessment of the effectiveness of their adaptation strategies in view of household (internal vulnerability) and institutional (external vulnerability) challenges is, in a direct way, questioning the viability of climate-smart agriculture practices so often lauded by some 'experts', South African rural development agencies and the provincial government as a panacea to climate impacts. Therefore, understanding local farmers' adaptation and resilience-building challenges, which are an additional burden given their historical livelihood precarity linked to chronic multidimensional poverty and resource limitations, is important not just for epistemological reasons but also for human development policy planning and programme implementation.

Smallholder Farmers' Adaptation Strategies in Southern Africa: A Synopsis

Climate adaptation and resilience is a growing body of knowledge in Southern Africa. Adaptation, with its multidimensional interpretations is

> adjustments in ecological, social, or economic systems in response to actual or expected climatic stimuli and their effects or impacts. It refers to changes

in processes, practices, and structures to moderate potential damages or to benefit from opportunities associated with climate change.

(Smit and Pilifosova 2001)

This characterization of adaptation extends the analysis beyond biophysical vulnerability "to the wider social and economic drivers of vulnerability and people's ability to respond" (Noble et al. 2014, 836) to climate shocks and stress. This is important because a holistic understanding of adaptation challenges faced by smallholder farmers in the Eastern Cape can only be complete when there is full appreciation of the convergence between socio-economic drivers and biophysical vulnerability. Relatedly, the confluence of these two influences farmers' adaptation options characterized hereto as "the array of strategies and measures available and appropriate to address adaptation needs" (Noble et al. 2014, 838). Adaptation options available to farmers can be *ex ante* or *ex post* climate impacts, but either way, they require multilayer implementation by farming households, communities, government entities and private institutions with a stake in resilience building. This means the adaptation process therefore involves various interrelated strategies by different actors directly and indirectly involved in the existential context of smallholder farmers.

The consciously planned or reactionary adaptation process, *ex ante* or *ex post* climate impacts, can result in effective adaptation or maladaptation. Effective adaptation, which is the extent to which adaptation strategies are achieving or have achieved intended objectives and closely attributable differential results, largely depends on adaptive capacity at two interconnected levels, systemic and individual levels. At the systemic level, "adaptive capacity is the ability of a system to adjust to climate change to moderate potential damages, to take advantage of opportunities, or to cope with the consequences" (McCarthy et al. 2001, 6). The ability of the system to adapt is partly mediated by the local ecosystem, its extent of biophysical vulnerability and the prevailing socio-economic environment. Relatedly, at the individual farmer level, who obviously exist within a particular human system, adaptive capacity "depends on such factors as wealth, technology, education, information, skills, infrastructure, access to resources, and management capabilities" (McCarthy et al. 2001, 8). Given that individual farmers are differently endowed with these resources, this means those with the least assets are the most vulnerable with lesser capacity to adapt and vice versa.

A number of studies show that many Southern African smallholder farmers dependent on rain-fed agriculture are adapting through shifting the onset of their planting season (Moyo and Dube 2014; Apraku et al. 2018) to align with early or delayed rainfall patterns. Relatedly, many farmers are adapting through changing their cropping choices from long season traditional maize production to short season drought-tolerant small grains (Call et al. 2019; Gars and Ward 2019; Phiri et al. 2020) and horticulture crops. However, for a variety of reasons, some smallholder farmers in Southern Africa have not embraced this climate-smart shift to small grain production. For example, in Zimbabwe, there is evidence that small grain production has declined over the last two decades (Gukurume 2013; Chanza 2018). This is partly due to the government's subsidization of maize seed and promotion of maize production as a pillar for food security under its climate-smart agriculture programme known locally as *Pfumvudza/Intwasa*. Relatedly, many farmers across Southern Africa overlook small grain production for various reasons. These include small grains' intensive production and processing labour demands, limited production knowledge and skills, low yields compared to maize, consumption preference of maize meal and limited government support for their production (Brazier 2015; Matthew 2015; Phiri et al. 2019).

In Southern African countries ravaged by incessant drought, some farmers with financial resources have invested in irrigation infrastructure or received the same as climate-smart development assistance from their governments or development partners. This is evident in South Africa where the government has significantly invested "in irrigation establishment, rehabilitation and revitalization" in an effort to "enhance food security and alleviate rural poverty" (Sinyolo et al. 2014, 145). Similarly, water harvesting and conservation (WHC) for smallholder agricultural production is gaining currency as an adaptation strategy in South Africa. There are numerous WHC techniques differing in scale, complexity and agricultural contexts. For example, in the Free State Province, homestead ponds or *matamo* are used to divert surface runoff "via small furrows and then stored for crop and domestic use" (Denison and Wotshela 2012, 11). In the Eastern Cape Province, "*gelesha* – the practice of turning the soil to increase winter rainfall infiltration…[is used] to ensure that any rain or dew or even frost infiltrates tilled soil thus increasing water availability for the next crop" (Denison and Wotshela 2012, 10). Although rudimentary, this water insecurity adaptation strategy is regularly used by farmers, at varying scales, during droughts and recurrent water crises.

The depletion of grazing pasture due to erratic rainfall patterns, high temperatures and droughts is also altering livestock production practices. In response to insufficient forage in their communal grazing areas, some farmers are practising transhumance (Moyo et al. 2013) in Botswana that involves temporary movement of animals to far-flung areas with better rangelands and water sources. Relatedly, many livestock farmers have recalibrated their animal husbandry focus to specialize in indigenous cattle breeds, for example, the Nguni breed in South Africa (Mapiye et al. 2009) and hard MaShona breed in Zimbabwe (Svotwa et al. 2007), both of which are adaptive to the heat and drought conditions of the Southern Africa savannah. Although diversification to small livestock (sheep and goats) is happening in countries that include Namibia (Kuiper and Meadows 2002), Malawi (Bie et al. 2008; Freeman et al. 2008), South Africa (Lehloenya et al. 2007) and Zimbabwe (Phiri et al. 2021), its uptake is not as high as anticipated by climate-smart livestock husbandry enthusiasts. One of the reasons for this reluctance by smallholder farmers to dispose of cattle is that they are stores of high economic value since they can be sold for significant cash and produce higher volumes of milk compared to goats and sheep traditionally seen as auxiliary livestock. Further, cattle are traditional sources of draught power for livelihood activities in Southern Africa (Casey and Maree 1993), and their ownership symbolizes high socio-cultural standing (Phiri et al. 2021) compared to sheep and goats ownership.

In addition to the above climate-smart agricultural practices, other studies have explored how "indigenous knowledge plays an important role in the ways local residents adapt to, and in some ways curb, the adverse impacts of climate change" (Apraku et al. 2018). Some have examined the extent to which rural households rely on local asset-based adaptation (Moyo 2017; Sinay and Carter 2020) and the nexus between adaptation and local socio-economic development (Le 2020). This paper builds on this asset-based scholarly trajectory through employing the Sustainable Livelihoods Framework (SLF) as its heuristic conceptual tool. In analysing adaptation and resilience challenges endured by Mqanduli smallholder farmers, it uses elements of the SLF. This is the ideal conceptual framework for such an analysis because the natural, physical, human, financial and social resources that constrain the adaptation of Mqanduli farmers are mirrored through variants of the SLF. While acknowledging the complexities of human existential conditions, the SLF argues that livelihood strategies and outcomes are shaped by the vulnerability context (shocks, trends and seasonality) and access to livelihood assets (human, natural, physical, financial and social) (Chambers and Conway 1992; Carney 1998; Scoones 1998). The impact of these assets on livelihoods is influenced by transforming structures (levels of government, private sector) and processes (laws, policies, culture, institutions) to determine livelihood strategies a household can diversify into to achieve different livelihood outcomes (reduced vulnerability, increased well-being, more income, improved food security) (Chambers and Conway 1992; Carney 1998; Scoones 1998). Notwithstanding its known limitations, the SLF best captures Mqanduli farmers' vulnerability context, the limited adaptation options at their disposal due to few assets, the influence of transforming structures and processes on their household assets and simultaneously on the resilience strategies they can pursue. The SLF is thus an appropriate analytical tool in this instance because while it acknowledges the agency of farmers, it is alive to how constraints emanating from local and national structures, institutions, trends and processes limit their resilience.

Research Setting and Methodological Approach

This study was conducted in King Sabata Dalindyebo (KSD) Municipality situated within the O.R. Tambo District in the Eastern Cape Province of South Africa. KSD Municipality is predominantly rural, has a population of 494,000, 117,623 households and geographically spans 3,072 km² which makes it the largest local municipality within the O.R. Tambo District (KSD Municipality 2022). The main rural settlement forms are medium and large privately owned commercial farms as well as communal smallholder farming households. While there are various economic activities in the rural hinterland that include forestry and tourism, subsistence agriculture is the main occupation for the majority (KSD Municipality 2022) in communal areas.

Within KSD, data collection was done in Mqanduli communal area, specifically in wards 21, 22, 23, 24 and 25. The primary reasons for purposively selecting these wards are, firstly, there is empirical evidence of recurrent droughts, high temperatures, veld fires and isolated flooding that affect the local environment, agriculture, infrastructure and livelihoods (KSD Municipality 2021). Secondly, the majority of the population in these five wards are predominantly smallholder farmers practising subsistence rain-fed agriculture perennially at risk of climate impacts. A mixed-methods approach was used to collect data from purposively selected respondents in the five wards. Two hundred and fifty household questionnaires, with closed- and open-ended questions, were completed. This number represents a 100% achievement of the target sample size. In addition, qualitative data was collected through 15 key informant interviews conducted with purposively selected agriculture extension officials from the provincial Department Agriculture and Rural Development, KSD rural and economic development officials, non-governmental organization programme officers, academics, community and traditional leaders as well as leaders of farmer groups. Field observations, pictures and transact walks were also done in the sampled wards enabling the collection of more qualitative data for enhanced critical analysis.

Adaptation Strategies for Resilience Building in Mqanduli, Eastern Cape

Livelihoods across the five study wards were predominantly ecosystem service based with the majority of respondents depending primarily on mixed farming, i.e., simultaneous crop production and animal husbandry. However, as shown in Table 9.1, households diversified their livelihood strategies beyond on-farm production to include synchronized reliance on social grants, remittances, informal economy income generation, small businesses, formal employment, precarious casual and self-employment.

This widespread reliance on a diversified livelihoods portfolio signifies household agency and ingenuity in communities beset by chronic vulnerability. One of the respondents (whose views mirror many others) justified this diversification as follows:

We have a family farm where we grow different crops that include maize, pumpkins, and watermelon. This is what we do every year with my wife and elder children. However, what we produce is not enough to last us until the

Table 9.1 Main Sources of Livelihood

Ward	Crop Farming	Livestock Production	Social Grant	Remittances	Informal Economy	Small Business	Casual Labour	Self-Employed	Formal Employment
21	42 (84%)	47 (94%)	33 (66%)	37 (74%)	11 (22%)	5 (10%)	13 (26%)	4 (8%)	3 (6%)
22	37 (74%)	41 (82%)	40 (80%)	40 (80%)	4 (8%)	2 (4%)	9 (18%)	8 (16%)	10 (20%)
23	40 (80%)	39 (78%)	38 (76%)	31 (62%)	9 (18%)	6 (12%)	14 (28%)	5 (10%)	13 (26%)
24	46 (92%)	45 (90%)	41 (82%)	43 (86%)	10 (20%)	8 (16%)	17 (34%)	9 (18%)	11 (22%)
25	39 (78%)	33 (66%)	35 (70%)	46 (92%)	7 (14%)	3 (6%)	11 (22%)	10 (20%)	15 (30%)

Source: Author's fieldwork data.

Table 9.2 Livestock Husbandry Adaptation Strategies

Adaptation Strategy	Number of Households	Percentage of Households
Fodder production for livestock	51	20
Feedlot for cattle	30	12
Change cattle breed to improve genetics	97	39
Switch from cattle to sheep farming	119	48

Source: Author's field data.

next harvest because in many years the rains are not enough and sometimes it is very hot just like now. Therefore, my two boys and myself we also do piece jobs like building huts for other people in surrounding villages then we use the money we earn to buy food and other basic things we need as a family. So it is a matter of us doing different things at the same time in order to survive as a family. You cannot do one thing here otherwise your family will suffer because farming only is not enough.

(Respondent 213, Ward 25)

This strategic decision to diversify strategies spreads risk and cushions each strategy's limitations in the resilience-building process. It also demonstrates the interdependence of various livelihood activities in household self-provisioning. In disaggregating these livelihood strategies, respondents were asked to indicate specific adaptation strategies their households introduced as a direct response to climate impacts. Households outlined the adaptation strategies as shown in Table 9.2.

The impact of droughts, high temperatures and uncontrolled communal grazing has depleted local rangelands and pasture-depriving livestock farmers a critical input in their production. In response, as shown in Table 9.2, the main adaptive practices are fodder production using local vegetation (n = 51, 20%) and feedlots (n = 30, 12%). This minority utilizing supplementary livestock feeding techniques signifies poor adoption of climate-resilient livestock production. The high cost of buying and hiring forage production equipment is the primary reason why many farmers do not own or rent it. As one of the livestock farmers, among many others, aptly put it:

I wanted to buy a hay bailer so that I could cut grass, bail it properly in preparation for feeding my cattle when the rains are gone and the veld is dry. I failed to buy it because when I went to East London and Port Elizabeth, the prices were too high for me. I even looked for a used one, second-hand one, but also the price was too much. So my cattle don't have adequate feed in winter and as a result last year three of them died because of hunger.

(Respondent 40, Ward 21)

Another reason that explains some farmers' low uptake of fodder production is the lack of relevant information. The majority of livestock farmers (n = 146) had no training, peer-to-peer or indigenous acquired knowledge of silage production. Consequently, farmers are not gathering, processing and producing forage, yet there is abundant local vegetation such as shrubs that can be processed into livestock forage with additional nutritional supplements. This failure to harness available natural resources (shrubs) for forage signifies inability to utilize locally available ecosystem services for resilience building.

Climate impacts have also pushed some farmers to recalibrate their animal husbandry. A minority 47% (n = 97) of livestock farmers have shifted from cross Brahman and Boran cattle breeds to those with drought-tolerant genetics, specifically the Nguni breed. However, a major limitation of this strategy is that while the Nguni cattle breed is drought and heat stress tolerant, compared to Brahman and Boran breeds, it has a small body frame, hence has lower live and carcass weight. Its average live weight price is +/−R10,000 in local markets. Such a low price means smallholder farmers make marginal profits when they decide to sell as an adaptation strategy or in cases where they require income for other livelihood needs. On the other hand, a majority of livestock farmers (n = 119, 58%) have completely shifted from cattle farming to small livestock husbandry specializing in local sheep breeds accustomed to the rangeland, erratic access to water and high temperatures. There are many advantages to this shift as explained by the local agricultural extension officer:

> it is encouraging to see many of our smallholder farmers shifting to sheep farming in response to current climate conditions. While we have played our part in encouraging them to do so, most of them did so on their own volition. For us, what is important is that their sheep are able to survive here where there is less water and grass. This is a good livestock farming strategy.
>
> (Agricultural Extension Officer 2, Department of Rural
> Development and Land Reform, Mqanduli)

This adaptation strategy, aligned to climate-smart livestock production, is indeed alive to current climate realities in the local area. It also demonstrates how the intersection of farmer agency and veterinary technical advice is a pathway towards effective adaptation. However, from an insider farmers' perspective, shifting to sheep production comes with socio-economic costs. Farmers argued that destocking a sizeable cattle herd and using the financial proceeds thereof to diversify into indigenous sheep farming is liquidation of high economic value inherent in cattle and an erosion of social prestige attached to ownership of a cattle herd. From an insider's perspective, this adaptation strategy thus results in inescapable decimation of their economic power and social standing in the community. This strategy is thus resulting in the reorganization of rural social structure whereby the traditionally patterned social arrangements and economic structural groupings based on levels of wealth (cattle ownership) are being redefined as households adapt. While this social reconfiguration might appear mundane to an outsider, the local reality is that responding to climate impacts is reorganizing societal strata and altering economic power dynamics as some livestock farmers' social standing is altered.

Further, for sheep farming to have sustainable impact, thus building resilience, it is important for smallholder farmers to transcend rudimentary animal breeding, management and veterinary practices. This requires investment in rams with superior genetics in order to build a breeding stock of ewes with larger body mass. Larger ewes will breed lambs that amass weight in a shorter period, thus fetching more at the point of sale. In addition, enhanced use of government-approved nutritional supplements and feed to improve animal health, meat texture and quality will lead to selling price increases for the benefit of the farmer. Relatedly, there is an urgent need to address farmers' limited knowledge of the sheep marketing value chain beyond the local and provincial market. The fact that a majority 52% (n = 107) of sheep farmers had never directly sold rams outside the province demonstrates they are yet to optimally profit from their stock. Instead, intermediaries immersed in the value chain buy locally at seemingly attractive prices for onward selling in other provinces and export markets with higher demand and better prices. Ultimately, the smallholder sheep farmers are the least beneficiaries in a value chain with potential to build and sustain their long-term resilience within the context of the climate crisis.

In addition to animal husbandry, crop farming remains central to local livelihoods. However, recurrent droughts and erratic rainfall patterns have perpetuated perennial decreases in cereal output especially maize, thus pushing some farmers (not the majority) with access to irrigation to adapt through diversifying into horticulture crops (butternuts, cabbages, tomatoes, potatoes and spinach) (Table 9.3).

The majority of horticulture farmers' testimonies resonated with this succinct explanation by one early adopter of this crop practice:

I had been growing maize for decades but my harvest kept on getting less and less every year. I even changed by maize seeds but that did not help because the rains were simply not enough and the sun was always scorching. So with financial assistance from my children, I drilled a well and bought irrigation equipment which you see here. I then started growing butternuts, onions, tomatoes and other vegetables instead of maize.

(Respondent 63, Ward 22)

Table 9.3 Crop Farming and Off-Farm Adaptation Strategies

Adaptation Strategy	*Number of Households*	*Percentage of Households*
Switch from maize to horticulture	63	31
Casual labour for income generation	50	20
Informal economy income generation	33	13
Abandon agriculture	17	7

Source: Author's field data.

These 'new' horticulture farmers justify shifting to cash crops based on their capacity to produce them at small-scale, marketability and affordability in local food markets. Their new dependence on irrigation production means they are on a path to build resilience to erratic rainfall patterns. However, this fundamental shift in the structure of farm production is not the norm because the majority of crop farmers remain maize producers since they have no access to irrigation facilities. Therefore, this means the capacity to diversify production systems and crops is intermediated by financial capital for irrigation. Those farmers without financial wherewithal, who happen to be the majority, are therefore unable to shift to horticulture and thus remain prone to reduced harvests and household food gaps.

Beyond on-farm adaptation, other household members (n = 83) are also diversifying into off-farm livelihood activities that include informal economy activities (n = 33) and seeking salaried casual labour (n = 50) opportunities in the rural service centre of Mqanduli and nearby town of Mthatha. While men and women of different ages were adapting in this manner, it was predominantly working age young men and young women of less than 40 years adapting off-farm. The need to diversify into the informal sector was explained by one of the young women traders as follows:

> I realized that we could no longer depend on farming as a family as we were harvesting less than we used to. I had to find something else to do away from my village and so I went to Mthatha where I do street trading. I sell traditional embroidery and traditional Xhosa attire there. The money that I make helps to support my family here.
>
> (Respondent 103, Ward 23)

This simultaneous utilization of on- and off-farm adaptation activities for livelihood diversification is not something new in rural Eastern Cape. Traditionally, rural livelihoods straddle the rural–urban divide, on-farm and off-farm activities. However, what we are witnessing now is an intensification of this diversification as households adapt to the new normal, the climate crisis.

In a few outlier cases, some households have completely abandoned subsistence agricultural production (n = 17) to concentrate on informal economy income generation and precarious casual labour. These off-farm adaptation strategies indicate the failure of on-farm responses to climate risk and shocks that have pushed some rural poor to diversify their livelihoods in geographic spaces beyond the rural hinterland. Farmers explain this complete abandonment of agricultural activities as an indicator of their inability to use different forms of meagre capital (human, financial and social) to adapt to climate impacts. This means their limited adaptation resources portfolio is the primary push factor behind migrating from the village rather than simple livelihood diversification. Their adaptation capacity and options have been exhausted, hence the drastic decision to make a living through non-agricultural activities. This, however, raises a sustainability question for the future. If a significant number were to adapt through diversifying into the informal economy and precarious casual labour, to what extent will these income-generating opportunities remain available, viable and profitable adaptation strategies? This cannot be answered now but is a line of enquiry worth exploring in future within the context of a struggling South African economy where the national unemployment rate remains high.

Adaptation Constraints and Farmer Responses

When conceptually analysed through the SLF lens, the foregoing findings demonstrate that adaptation challenges endured by smallholder farmers in Mqanduli are primarily a product of, among other factors, their biophysical vulnerability and limited direct access to physical, human, financial and social assets. Despite their agency, they remain vulnerable because the "vulnerability of a community or society, which is the interaction between three main dimensions: exposure, sensitivity, and adaptive capacity, is affected by both weather and climate extremes and non-climatic factors" (Le 2020, 741). Therefore, their perpetual vulnerability to climate impacts is a function of their inability to adapt effectively due to asset constraints. The findings above affirm how their limited assets (different types of capital), mediated by transforming structures and processes, curtails adaptation; hence, climate shocks, stresses and trends continue to wreak havoc to their livelihoods and existential conditions unabated.

Although adaptation constraints differ across households, depending on access to assets and the influence of external socio-economic and environmental factors, there are some common community-level barriers. For example, limited access to physical capital in the form of climate variability information is a common operational challenge hampering resilience building. While basic daily weather reports are broadcast in local languages on local radio (e.g., *Umhlobo Wenene* FM) and national television channels (e.g., South African Broadcasting TV channels 1, 2 and 3), these are not entirely useful for local farming decision-making. This is because they are not specific to their farming locality since they are generalized to the nearest town and not longitudinal in their projection. Relatedly, climate variability data is also not easily accessible by local smallholder farmers. This lack of access to climate information means decisions guiding onset of the planting season, suitable crop variety and duration of farming season are arbitrary. Without adequate and requisite climate variability and change information, building resilience thus remains elusive for the majority of smallholder farmers.

Further, the inability of rural development mandated provincial government departments, specifically the Department of Rural Development and Agrarian Reform (DRDAR) and Eastern Cape Rural Development Agency (ECRDA), to share climate variability and change data with smallholder farmers in a language, communication format and platform amenable to them partly accounts for their adaptation challenges. While agriculture extension officers from DRDAR confirmed undergoing training on basic climate change mitigation, adaptation and resilience, this was not customized to the local agrarian context, hence remains of little value in their advisory information-sharing sessions with farmers. The inadequacies of their knowledge and communication approach are thus affecting the utility of their extension work. As a result, smallholder farmers deprived access to this physical capital (information) are not fully informed about many climate-resilient agricultural practices suitable for their agro-ecological region, hence their continued adaptation gaps.

Alongside the climate information deficit, limited access to climate finance (financial capital) is directly incapacitating farmers' agency and adaptation actions. Since household financial wherewithal and levels of exposure varied, this means their ability to adapt and attendant challenges differed. More financially and materially resourced households, who were the minority, had more choices, hence faced less barriers towards resilience. The majority did not have adequate resources for effective and sustainable adaptation. Not only were they historically resourced less, but climate impacts further eroded their financial and material holdings, deepening exposure and vulnerability. This aggravated their inability to effectively adapt and build resilience. These challenges were exacerbated by the failure of DRDAR and other rural development-oriented government departments to inject substantial climate finance to supplement farmers' resilience-building strategies. The meagre climate finance for livestock supplementary feed (only for livestock feedlot members) and subsidized fertilizer for selected crop farmers is testament of the provincial government's failure to augment the agency of farmers. Without adequate climate finance, farmers could not procure forage-making equipment nor process local vegetation to produce silage for lean season livestock supplementary feeding. However, this provincial government climate resilience-programming shortcoming is surmountable through rethinking agriculture-funding priorities. Instead of continuing to channel funds in traditional agricultural production activities without a sense of climate change realities, relevant rural development government departments should mainstream climate finance in their programming in order to build a resource base for climate-resilient rural development.

While climate finance and technical support from state institutions are important elements for climate-resilient development, the agency of farmers (as individuals or groups) is the foundation of climate action. A quaint approach of understanding their agency is through exploring whether their social capital mediates adaptation struggles to the climate crisis. There is evidence that the farmers regularly rely on social resources accessed through their horizontal social networks. Illustrative cases include peer-to-peer information sharing on livestock breeding techniques, veterinary, communal rangeland management and livestock market prices. Similarly, information on crop input prices, seed varieties, horticulture practices, food market dynamics and available government support programmes filtered through farmer groups. These farmer networks are thus social capital that facilitates their cooperation for individual and group benefit. However, social capital's efficacy in aiding resilience was inhibited by farmers' human and financial capital constraints. For example, limited human capital in the form of climate variability data and knowledge meant resilience-building information circulated in their networks was of limited scope. Specialized peer-to-peer social learning among fellow farmers was thus hamstrung by knowledge limitations. Simultaneously, limited financial capital within farmer networks meant sharing material resources such as transhumance of livestock (previously a common traditional practice) was not possible in a community with largely limited economic means.

Lastly, the provincial government's misplaced climate programming priorities also partly account for adaptation gaps experienced by farmers. At a conceptual level, this inappropriate programming by DRDAR exemplifies the influence of transforming structures and processes (as articulated in the SLF) on farmers' adaptive capacity. At a practical agricultural production level, some of the climate resilience programmes encouraged in the study community are dysfunctional on so many accounts. Firstly, although DRDAR agricultural extension officers promote climate adaptation strategies in the farming community, some of them are not tailor-made for local resilience-building priorities. For example, their promotion of adapting through switching from maize to small grains, although climate-smart in other regions, is disconnected from local farming realities since locals evidently prefer horticulture instead of small grains. Secondly, the climate variability data they sporadically share with farmers has general provincial scope, hence of little utility in guiding local adaptation decision-making. Consequently, farmers' resilience-building decision-making is arbitrary, not informed by locally relevant climate data and appropriate strategies. The omissions of these provincial government programming can, however, be easily addressed. Agriculture extension officers (from DRDAR) should simply revisit their climate resilience-building content to ensure that climate data and adaptation strategies promoted among farmers are practical at the local scale, relevant and sustainable.

Conclusion

The agency of smallholder farmers in the Eastern Cape as they adapt to climate impacts demonstrates their ingenuity and endogenous ability to use ecosystem services and meagre socio-economic resources as they attempt to build resilience. However, their agency in pursuit of effective adaptation is mediated by complex interactions between underlying vulnerability drivers (i.e., access to physical, human, financial and social capital) and biophysical vulnerability (natural capital). These factors influence choice of adaptation strategies, their effectiveness and the extent to which they assuage climate impacts. Further, these underlying and biophysical vulnerability factors are the genesis of observed adaptation challenges. With variations across farmer households, access (or lack thereof) to different forms of capital determines the character and extent of challenges endured as well as farmers' ability to ameliorate these constraints. This is personified by how limited financial capital is partly responsible for the majority of farmers' inability to enhance adaptation through irrigated horticulture, procurement of livestock with drought-tolerant genetics and failure to buy silage equipment for fodder production. Similarly, their human capital constraints partly explain why many cannot access locally relevant climate variability data that consequentially leads to arbitrary decision-making in adaptation crop choices and farming practices. Relatedly, farmer groups' social capital constraints mean resilience-building information shared within their social networks is of limited scope and utility, thus constraining effectiveness of their adaptation strategies. Ultimately, these adaptation challenges are pushing some rural dwellers to abandon subsistence agriculture and migrate to nearby towns in search of income-generating opportunities in the informal sector and precarious casual labour sector. Even though this migration does not signify an overall reconfiguration of rural livelihoods, it indicates how climate impacts are deepening rural existential struggles resulting in rural to urban migration for survival.

However, given their endogenous agency and ingenuity, there is policy and programming room to address these challenges and enhance farmers' adaptive capacity. Since Eastern Cape provincial government rural development and agrarian reform departments are already investing material and technical resources in rural production systems, what is required is a readjustment that centralizes the climate crisis in their programming. This entails actively mainstreaming climate adaptation and resilience building in their programmes. Through this, smallholder farmers will have an opportunity to enhance their adaptive capacity by benefiting from state-supported, customized climate finance, climate-smart agriculture skills enhancement, access to locally relevant climate variability data, increased uptake of resilient crop and livestock farming technologies. With the support of a state-aligned, consistent and structured financial, technical, informational and capacity-building intervention, there will be sufficient practical ground for smallholder farmers to ameliorate adaptation challenges currently at the epicentre of their precarious existence.

References

Apraku, A., W. Akpan and P. Moyo. 2018. "Indigenous Knowledge, Global Ignorance? Insights from an Eastern Cape Climate Change Study." *South African Review of Sociology* 49 (2): 1–21. https://doi.org/10.1080/21528586.2018.1532813.

Bie, S.W., D. Mkwambisi and M. Gomani. 2008. "Climate Change and Rural Livelihoods in Malawi." *Noragric Report* 41. Department of International Environment and Development Studies. Oslo: Norwegian University of Life Sciences.

Brazier, A. 2015. *Climate Change in Zimbabwe: Facts for Planners and Decision Makers.* Harare: Konrad Adenauer Stiftung.

Burls, N.J., R.C. Blamey, B.A. Cash, E.T. Swenson, A. al Fahad, M.J.M. Bopape, D.M. Straus and C.J.C. Reason. 2019. "The Cape Town 'Day Zero' Drought and Hadley Cell Expansion." *NPJ Climate and Atmospheric Science* 2. https://doi.org/10.1038/s41612-019-0084-6.

Call, M., C. Gray and P. Jagger. 2019. "Smallholder Responses to Climate Anomalies in Rural Uganda." *World Development* 155: 132–144. https://doi.org/10.1016/j.worlddev.2018.11.009.

Carney, D. 1998. "Implementing the Sustainable Rural Livelihoods Approach." In *Sustainable Rural Livelihoods: What Contributions Can We Make?*, edited by D. Carney. London: Department for International Development: 3–23.

Casey, N.H. and C. Maree. 1993. *Livestock Production Systems: Principles and Practice.* Brooklyn: Agri Development Foundation.

Chambers, R. and G.R. Conway. 1992. "Sustainable Rural Livelihoods: Practical Concepts for the 21st Century." *IDS Discussion Paper* 296. Brighton: Institute of Development Studies.

Chanza, N. 2018. "Limits to Climate Change Adaptation in Zimbabwe: Insights, Experiences and Lessons." In *Limits to Climate Change Adaptation: Climate Change Management*, edited by W.L. Filho and J. Nalau. Cham: Springer. https://doi.org/10.1007/978-3-319-64599-5_6.

Chersich, M.F., C.Y. Wright, F. Venter, H. Rees, F. Scorgie and B. Erasmus. 2018. "Impacts of Climate Change on Health and Wellbeing in South Africa." *International Journal of Environmental Research and Public Health* 15 (1884). https://doi.org/10.3390/ijerph15091884.

Denison, J.A. and L. Wotshela. 2012. "An Overview of Indigenous, Indigenised and Contemporary Water Harvesting and Conservation Practices in South Africa." *Irrigation and Drainage* 61 (2): 7–23. https://doi.org/10.1002/ird.1689.

Department of Cooperative Governance and Traditional Affairs. 2019. "Update on Relief for Families Affected by the Floods Disasters in Eastern Cape and KZN." http://www.cogta.gov.za/?p=6669.

Dube, K., G. Nhamo and D. Chikodzi. 2021. "Flooding Trends and Their Impacts on Coastal Communities of Western Cape Province, South Africa." *GeoJournal*. https://doi.org/10.1007/s10708-021-10460-z.

Engelbrecht, F.A. and P.M.S. Monteiro. 2021. "The IPCC Assessment Report Six Working Group 1 Report and Southern Africa: Reasons to Take Action." *South African Journal of Science* 117 (11/12): 12–18.

Freeman, H.A., S. Kaitibie, S. Moyo and B.D. Perry. 2008. "Livestock, Livelihoods and Vulnerability in Lesotho, Malawi and Zambia: Designing Livestock Interventions for Emergency Situations." *Research Report* 8. Nairobi: International Livestock Research.

Gars, J. and P.S. Ward. 2019. "Can Differences in Individual Learning Explain Patterns of Technology Adoption? Evidence on Heterogeneous Learning Patterns and Hybrid Rice Adoption in Bihar, India." *World Development* 115: 178–189. https://doi.org/10.1016/j.worlddev.2018.11.014.

Gukurume, S. 2013. "Climate Change, Variability and Sustainable Agriculture in Zimbabwe's Rural Communities." *Russian. Journal of Agricultural and Socio-Economic Sciences* 2 (14): 89–93.

King Sabata Dalindyebo Municipality. 2021. "King Sabata Dalindyebo Local Municipality IDP Review 2020/21." https://www.cogta.gov.za/cgta_2016/wp-content/uploads/2020/11/KING-SABATA-DALINDYEBO-Final-IDP-2020-2021.pdf.

King Sabata Dalindyebo Municipality. 2022. "About." https://ksd.gov.za/about/.

Kuiper, S.M. and M.E. Meadows. 2002. "Sustainability of Livestock Farming in the Communal Lands of Southern Namibia." *Land Degradation and Development* 13 (1–15). https://doi.org/10.1002/ldr.476.

Le, T.D.N. 2020. "Climate Change Adaptation in Coastal Cities of Developing Countries: Characterizing Types of Vulnerability and Adaptation Options." *Mitigation and Adaptation Strategies for Global Change* 25: 739–761. https://doi.org/10.1007/s11027-019-09888-z.

Lehloenya, K.C., J.P.C. Greyling and L.M.J. Schwalbach. 2007. "Small-Scale Livestock Farmers in the Peri-urban Areas of Bloemfontein, South Africa." *South African Journal of Agricultural Extension* 36: 217–228.

Livingston, G., S. Schonberger and S. Delaney. 2011. "Sub-Saharan Africa: The State of Smallholders in Agriculture." Paper Presented at the IFAD Conference on New Directions for Smallholder Agriculture, 24–25 January 2011. Rome: International Fund for Agricultural Development.

Mapiye, C., M. Chimonyo, K. Dzama, J.G. Raats and M. Mapekula. 2009. "Opportunities for Improving Nguni Cattle Production in the Smallholder Farming Systems of South Africa." *Livestock Science* 124 (1–3): 196–204.

Matthew, S. 2015. "The Feasibility of Small Grains as an Adoptive Strategy to Climate Change." *Russian Journal of Agricultural and Socio-Economic Sciences* 4 (15): 40–55.

McCarthy, J.J., O.F. Canziani, N.A. Leary, D.J. Dokken and K.S. White. 2001. *Climate Change 2001: Impacts, Adaptation, and Vulnerability*. Cambridge: Cambridge University Press.

Mkhize, M. 2022. "KwaZulu Natal Households to Receive R15.4m in Aid after Rain and Flood Damage." https://www.timeslive.co.za/news/south-africa/2022-01-06-kzn-households-to-receive-r154m-in-aid-after-rain-and-flood-damage/.

Moser, C., A. Norton, A. Stein and S. Georgieva. 2010. *Pro-Poor Adaptation to Climate Change in Urban Centers: Case Studies of Vulnerability and Resilience in Kenya and Nicaragua*. Washington, DC: World Bank.

Moyo, P. 2017. "Vulnerability and Assets Nexus in Climate Change Adaptation: Reflections from South Africa and Zimbabwe." In *Revisiting Environmental and Natural Resource Questions in Sub-Saharan Africa*, edited by W. Akpan and P. Moyo, 75–91. Newcastle: Cambridge Scholars Publishing.

Moyo, P. and T. Dube. 2014. "Edging Towards a Tipping Point? An Appraisal of the Evolution of Livelihoods under Climate Change in Semi-arid Matobo, Zimbabwe." *International Journal of Development and Sustainability* 3 (6): 1340–1353.

Moyo, B., S. Dube and P. Moyo. 2013. "Rangeland Management and Drought Coping Strategies for Livestock Farmers in the Semi-arid Savanna Communal Areas of Zimbabwe." *Journal of Human Ecology* 44 (1): 9–21.

Noble, I.R., S. Huq, Y.A. Anokhin, J. Carmin, D. Goudou, F.P. Lansigan, B. Osman-Elasha and A. Villamizar. 2014. "Adaptation Needs and Options." In *Climate Change 2014: Impacts, Adaptation, and Vulnerability.* Part A: Global and Sectoral Aspects. Contribution of Working Group II to the Fifth Assessment Report of the Intergovernmental Panel on Climate Change, edited by C.B. Field, V.R. Barros, D.J. Dokken, K.J. Mach, M.D. Mastrandrea, T.E. Bilir, M. Chatterjee, K.L. Ebi, Y.O. Estrada, R.C. Genova, B. Girma, E.S. Kissel, A.N. Levy, S. MacCracken, P.R. Mastrandrea and L.L. White, 833–868. Cambridge: Cambridge University Press.

Nyoni, N.M.B., S. Grab, E. Archer and J. Malherbe. 2021. "Temperature and Relative Humidity Trends in the Northernmost Region of South Africa, 1950–2016." *South African Journal of Science* 117 (11/12). https://doi.org/10.17159/sajs.2021/7852.

Otto, F.E.L., P. Wolski, F. Lehner, C. Tebaldi, G.J. van Oldenborgh, S. Hogesteeger, R. Singh, P. Holden, N.S. Fučkar, R.C. Odulami and M. New. 2018. "Anthropogenic Influence on the Drivers of the Western Cape Drought 2015–2017." *Environmental Research Letters* 13 (12). https://doi.org/10.1088/1748-9326/aae9f9.

Phiri, K., S. Ndlovu, M. Mpofu and P. Moyo. 2020. "Climate Change Adaptation and Resilience Building through Small Grains Production in Tsholotsho, Zimbabwe." In *Handbook of Climate Change Management*, edited by W.L. Filho., J.M. Luetz and D. Ayal. Cham: Springer. https://doi.org/10.1007/978-3-030-57281-5_1.

Phiri, K., S. Ndlovu, M. Mpofu, P. Moyo and H.C. Evans. 2021. "Addressing Climate Change Vulnerability through Small Livestock Rearing in Matobo, Zimbabwe." In *African Handbook of Climate Change Adaptation*, edited by W.L. Filho, N. Oguge, D. Ayal, L. Adeleke and I. da Silva, 639–658. Cham: Springer. https://doi.org/10.1007/978-3-030-45106-6_121.

Phiri, K., T. Dube, P. Moyo, C. Ncube, S. Ndlovu and G. Buchenrieder. 2019. "Small Grains "Resistance"? Making Sense of Zimbabwean Smallholder Farmers' Cropping Choices and Patterns within a Climate Change Context." *Cogent Social Sciences* 5 (1). https://doi.org/10.1080/23311886.2019.1622485.

Schulze, R.E. 2016. "On Observations, Climate Challenges, the South African Agriculture Sector and Considerations for an Adaptation Handbook." In *Handbook on Adaptation to Climate Change for Farmers, Officials and Others in the Agricultural Sector of South Africa*, Technical Report, edited by R.E. Schulze, 1–17. Pretoria: Department of Agriculture, Forestry and Fisheries.

Scoones, I. 1998. "Sustainable Rural Livelihoods: A Framework for Analysis." *IDS Working Paper* 72. http://www.ids.ac.uk/ids/bookshop/wp/wp72.pdf.

Sinay, L. and R.W. Carter. 2020. "Climate Change Adaptation Options for Coastal Communities and Local Governments." *Climate* 8 (1). https://doi.org/10.3390/cli8010007.

Sinyolo, S., M. Mudhara and E. Wale. 2014. "The Impact of Smallholder Irrigation on Household Welfare: The Case of Tugela Ferry Irrigation Scheme in KwaZulu-Natal, South Africa." *Water SA* 40 (1): 145–156. http://doi.org/10.4314/wsa.v40i1.18.

Smit, B. and O. Pilifosova. 2001. "Adaptation to Climate Change in the Context of Sustainable Development and Equity." In *Climate Change 2001: Impacts, Adaptation, and Vulnerability*, edited by J.J. McCarthy, O.F. Canziani, N.A. Leary, D.J. Dokken and K.S. White, 879–912. Cambridge: Cambridge University Press.

Svotwa, E., H. Hamudikuwanda and A. Makarau. 2007. "Influence of Climate and Weather on Cattle Production in Semi-Arid Communal Areas of Zimbabwe." *Electronic Journal of Environmental, Agricultural and Food Chemistry* 6 (2): 1838–1850.

Tshiala, M.F., J.M. Olwoch and F.A. Engelbrecht. 2011. "Analysis of Temperature Trends over Limpopo Province, South Africa." *Journal of Geography and Geology* 3 (1). https://doi.org/10.5539/jgg.v3n1p13.

Zikhali, Z.F., P.L. Mafongoya, M. Mudhara and O. Jiri. 2020. "Climate Change Mainstreaming in Extension Agents Training Curricula: A Case of Mopani and Vhembe District, Limpopo Province, South Africa." *Journal of Asian and African Studies* 55 (1): 44–57.

10 Just Adaptation and Climate Justice

Possibilities for Youth Development in Rural South Africa

Tanaka C. Mugabe

Introduction

The saying that the future belongs to the youth poses a conundrum in South Africa as to what that future entails when faced with rising temperatures and the damage that centuries of fossil fuels and other emissions have had on developmental prospects owing to the current and future effects of climate change (United Nations, 2020). Apart from the ever-increasing climate challenges, South African youth are compounded with socio-economic problems ranging from high unemployment rates which was standing at an alarming 59.6% in the third quarter of 2022 (Trading Economics, 2023), poverty, crime, HIV and AIDS, there is also alcohol and drug abuse, absence of parents in families, unplanned pregnancies and homelessness. While the national government continuously endeavours to plan for the developmental aptitude for youth, these developments have had limited to no input to the reality of the integrated biophysical and social environment that poses a risk for the survival of humanity as we know it. The only way of conceiving current socio-ecological problems is by framing them in terms of an environmental crisis which could, hypothetically, be solved by the very same societal model that created it. This is noting that South African youth are a troubled generation. The problems they face hamper efforts pertaining to youth development and empowerment and pose challenges in the attempt for climate change projects, notably those identified to enable climate justice. This chapter therefore analyses the possibilities for climate justice with South African youth through unpacking their socio-economic challenges and how social justice can be incorporated into adaptation planning.

The emergence of climate justice and just adaptation was given an impetus by the rise of climate change on the globe. South Africa, in particular the Centre for Environmental Rights (Scholes & Engelbrecht, 2021), noted that its contribution to climate change is mainly through greenhouse gas (GHG) emissions. South Africa's GHG emissions is 1.1% of the global emissions, whereas the country's share of global Gross Domestic Product (GDP) is only 0.6% (University of Cape Town, 2017).These GHG emissions are cumulatively human driven, also referred to as "anthropogenic" emissions resulting from the combustion of fossil fuels (coal, oil and natural gas). Most of Southern Africa depends on coal, oil and gas as sources of energy, and any change impacts on the social and economic landscape of society

DOI: 10.4324/9781003397120-12

and poses questions on the implications of human rights and vulnerable groups within the community.

The effects and impact of climate change have taken a toll on South Africa in 2022. The heavy rainfalls that occurred in parts of KwaZulu Natal (province in South Africa) caused enormous damage to the public infrastructure such as roads, rail, bridges, schools and properties. Many businesses including shipping and logistic companies saw their containers being washed away by the floods. These floods disrupted electricity and water supply in many areas in the province, with the eThekwini metro being the most affected area in this regard. While governments and business rethink restructuring and reconstruction, it is imperative for the social scientific community to unpack not only the intrinsic causes of such patterns which are indicative of effects of climate change but also just measures that have detrimental effect of distraction on a socially challenged community, more so rural communities, which is incomprehensible.

Thus, the avenues of climate justice must be considered within the context of the unique qualities that include countries' developmental status, environmental patterns, social dynamics and economic imperatives such as its natural resource dependence. The approach proposed in this chapter is a systems thinking for just adaptation as a measure of climate justice. Therefore, three concepts are brought to the fore, namely, climate justice, just adaptation and systems. These concepts will be unpacked and engaged within the context of youth, using rural South Africa as a case study.

Climate Justice Discourse

Climate change is undoubtedly a global phenomenon, and activists and policy advocates have indicated that addressing the effects requires just mechanisms that consider the human element giving rise to two key concepts, climate justice and just adaptation, with the former gaining much momentum in the past decade. Pursuing climate justice means combatting social injustice, gender injustice, economic injustice, intergenerational injustice and environmental injustice. The intersectionality of these challenges must be acknowledged in order to address them holistically. For example, some climate projects inadvertently create climate injustices when local communities are displaced for a conservation or renewable energy initiative. In addition, the climate projects are often in isolation of recognising social dynamics such as the escalation of behavioural degradation of youth particularly in South Africa. It is through understanding the predicament of youth within their context can avenues for redress be applied and more so apply a systems thinking for possibilities of harnessing a developmental trajectory for youth and combatting climate woes through an integrated climate justice approach.

The climate crisis we are currently experiencing is as a result of a system which prioritises profit over sustainability. This is the best time to reimagine a systems transformation for climate justice. Climate justice requires approaches that address the unequal burdens in vulnerable communities and groups. This implies climate justice ensures linking human rights with development and climate action. Development cannot be delinked from climate action and vice versa. Throughout, a human rights-based approach is necessary. For example, with the rapid pace of urbanisation, a rights-based approach is crucial for addressing water, sanitation and health challenges which are exacerbated by climate change in the most vulnerable communities.

Vulnerable communities require a people-centred approach to climate justice. This entails ensuring representation, inclusion and protection of the rights of those most vulnerable to the effects of climate change. Africa and more so some parts of Southern Africa by their nature are considered to have more vulnerable communities. Southern Africa is particularly vulnerable to climate change because of its geographical location and socio-economic development state. It is an already warm and dry region. Freshwater availability is already critically limited in Southern Africa and will be reduced in future as a result of decreasing rainfall and increasing evaporation. These impacts will amplify as the level of global warming increases. There is a high likelihood that agricultural production in Southern Africa, including staple crops and livestock, will be reduced relative to the no-climate change case. This is because the region is already beyond the temperature optimum for most crop and livestock production, and crop and forage production in an already dry country decreases if soil moisture decreases further. The number, intensity and duration of heat waves in South Africa will increase steeply in future as a result of global warming. The capacity to perform manual labour out of doors decreases dramatically as the occurrence of heat waves increases. The risk of severe storms, including intense tropical cyclones and very intense thunderstorms, increases with climate change in Southern Africa. As a result, loss of life, injury and damage to infrastructure also increase (Scholes and Engelbrecht, 2021).

Also, Southern Africa is already a warm, sub-tropical and semi-arid region and researchers are convinced that they will be harder hit than temperate regions, and South Africa, despite being the most developed country in the Southern African Development Community (SADC), will suffer enormous negative physical, socio-economic and ecological impacts, under all scenarios. These will include extreme heat stress, extreme weather events, including storms, flooding and droughts, sea-level rise and coastal damage, crop failures and food insecurity, water stress, disease outbreaks, various forms of economic collapse and social conflict and mass migration to informal settlements around urban areas (King, 2021).

On a global front, the Paris Agreement (United Nations, 2015) has been instrumental to tackle climate change and its negative impacts that are spreading throughout the globe. The Paris Agreement has been used as the guide for nations to ascertain plans for climate justice and just adaptation. South Africa has also used the agreement through its enabling Nationally Determined Contribution, or NDC, to address the effects of climate change through a national climate action plan; however, the aspect of climate justice and just adaptation or transition has not received widespread influence into local planning and policies. The identified long-term goals for the Paris Agreement to all nations include substantially reduce global GHG emissions to limit the global temperature increase in this century to 2°C while pursuing efforts to limit the increase even further to 1.5°C, review countries' commitments every five years, provide financing to developing countries to mitigate climate change, strengthen resilience and enhance abilities to adapt to climate impacts.

In ensuring South Africa succeeds with reducing global gas emissions, South Africa adopted the strategy Low-Emission Development Strategy (SA LEDS) 2050 in 2004 which aims to integrate socio-economic development and reduce emissions. South Africa has implemented a comprehensive set of strategies, policies and sector plans within key sectors of the economy. These include, among others, Integrated Resource Plan (IRP), Energy Efficiency Strategy, the Industrial Policy Action Plan (IPAP), the Green Transport Strategy (GTS), the Climate Change Adaptation and Mitigation Plan (CCAMP) for the South African Agricultural and Forestry Sectors and National Waste Management Strategy (NWMS). Together, these policies should drive far-reaching change throughout the economy and society and further the country's low-carbon development agenda (Department of Forestry Fisheries and Environment, 2018). Various studies exist to indicate the milestones of each of the strategies within their specific sector and impact thereof. However, limited data exists on the comprehensive contribution across the different subsectors.

As indicated above, the Paris Agreement works on a five-year cycle of increasingly ambitious climate action carried out by countries. Every five years, each country is expected to submit an updated national climate action plan that is the NDC. To date, South Africa will contribute their NDC in 2023 as prescribed. On the other hand, on financing, South Africa has benefited adaptation funding organized through the UNFCCC Conference of the Parties, more commonly referred to as COP, through the South Africa National Biodiversity Institution which is the country's implementing agency. Two projects were funded which were both for rural communities in Limpopo and KwaZulu Natal provinces, respectively. The funding in KwaZulu Natal was towards a vulnerable rural community to conduct projects within a rural district of uMgungundlovu District Municipality (UMDM), KwaZulu Natal Province, South Africa. The project was aimed at reducing climate vulnerability and increasing the resilience and adaptive capacity in rural and peri-urban settlements of small-scale farmers in productive landscapes that are threatened by climate variability and change, through an integrated adaptation approach. This was enabled through implementing a suite of complementary gender-sensitive project interventions, focusing on early warning and ward-based disaster response systems, ecological and engineering infrastructure solutions specifically focused on vulnerable communities, including women, integrating the use of climate-resilient crops and climate-smart techniques into new and existing farming systems and disseminating adaptation lessons learned and policy recommendations to facilitate scaling up and replication.

Nonetheless, the effects of climate change continuously rampage the district. In 2022, a series of floods were reported. The heavy rainfalls that occurred in parts of KwaZulu Natal (province in South Africa) caused enormous damage to the public infrastructure such as roads, rail, bridges, schools and properties. Many businesses including shipping and logistic companies saw their containers being washed away by the floods. These floods disrupted electricity and water supply in many areas in the province, with the eThekwini metro being the most affected area in this regard. While governments and businesses rethink restructuring and reconstruction, it is imperative for the social scientific community to unpack the intrinsic causes of such patterns which are indicative of effects of climate change.

The detrimental effect of distraction on a socially challenged community, more so rural communities, is incomprehensible; therefore, climate justice to enhance adaptation in recognising that the problem is here to stay is required. The limitation on the subject areas that were funded through the adaptation fund is a gap in understanding the social fibre and the reality of the local people, their environment and how it manifests within a climate change conundrum. This chapter will later provide a case study to indicate some of the empirical variables that such projects should seek to integrate in their adaptation planning.

Climate Justice Theory Implications

Climate justice theory emerged to enable a more determined focus to what most policy and environmental advocates previously addressed through broader environmental justice. Schlosberg (2012) has since suggested that climate justice theories can be engaged within various facets that enable looking at climate not only through the lens of the physical but also understanding of the political aspects that determine distribution of goods and risks as well as the social and cultural bearings within the context that can either influence or hinder climate action. To this end, the focus until now has been on a capability approach which suggests the understanding of the relationship between human beings and the nonhuman world (Schlosberg, 2012). However, the conversation of climate justice tends to lose momentum when not allowed to manifest within the existing systems and contexts of the environment. This predicament is not only existing in South Africa but also in the majority of developing countries.

Achieving climate justice means understanding that not everyone has contributed to climate change in the same way. While everyone must do their part to address climate change, the burden should not be borne by those who have contributed the least. The world's richest 10% are responsible for 50% of GHG emissions, and the poorest 50% are only responsible for 10% despite population and energy consumption increasing.

A concerted effort to integrate agendas through a climate justice praxis and policy, more so local policy, is required if the benefits are to be experienced widely. The concept of justice has given rise to approaches of just transition and the prior concept of just adaptation, while some incline to just transition; rightfully so, this chapter opts to explore just adaptation as previously and successfully applied in 2012 by the United Kingdom (Brisley, 2012) and more recently reported by Australia in 2022. The approach to just adaptation incorporates both theory and praxis.

Unpacking Just Adaptation

Just adaptation emerges from environmental justice and critical development perspectives. Rawls (1971) defined it as the fair allocation of material and social benefits among people over space and time. The concept has evolved over time to negate its negative association with individualism and goes beyond the fair distribution of goods and services and fair procedures to take in Amartya Sen's notion of capabilities, that is, to exert political choices over the use of those resources. The large corpus of scholarship within just adaptation is concerned with the discursive and allocative mechanisms that drive the production of justices and injustices to help explain how environmental and development priorities of equity and justice stand in stark contrast to local experiences of inequity and injustice (See & Wilmsen, 2020). Therefore, just adaptation enables an analysis into how adaptation policies take into account social justice issues particularly in vulnerable groups such as the South African youth used in this context.

It will be recalled that the concept of adaptation has had various contested views for its meaning and application. Adaptation in the climate justice and climate change literature emerged through the United Nations about 25 years ago (in 2001). The idea of adaptation has continued to evolve within policies and strategies including the Paris Agreement (Eisenack & Stecker, 2012). The Intergovernmental Panel on Climate Change (Intergovernmental Panel on Climate Change, 2001) initially provided the definition of adaptation to mean, "adjustment in natural or human systems in response to actual or expected climatic stimuli or their effects, which moderates harm or exploits beneficial opportunities". Adaptation conceptual framing within climate justice can be targeted at changing contextual conditions or at reducing damage. Therefore, adaptation cannot change vulnerability; however, in the long run, actual adaptation reduces vulnerability and, consequently increases the potential to adapt (to integrate social and biophysical vulnerability by adopting the perspective of coupled social-ecological systems). Firstly, a distinction is established between meaning of adaptive capacity, which relates to potential to reduce social vulnerability through realisation of those environmental factors for adaptation that in turn decreases biophysical vulnerability wherein another relates to adaptative capacity relates to the potential for adaptation within systems. The application of adaptive capacity is thus determined by the scenario in question.

With our social imagination, adaptation is related precisely to singular actions that are exercised by actors. This implies the understanding of how a person or persons within communities, particularly vulnerable context, act or harness adaptive capacity requires understanding of the nature and structure of the respective systems in which they function. This is noting that adaptation is a response to climate change that can be either an environmental change or human action. To provide a more intrinsic explanation to the human action, I refer to Parsons' meaning of adaptation conceived through a four-subsystem model of the social environment around four "tasks" facing a social system in relation to its environment. These four subsystems were Goal Attainment (the polity), Adaptation (the economy), Integration (cultural system of general values which is concerned with law and social control) and Latency (the normative problem of motivation to fulfil positions in the social system) (the GAIL system). There are definite problems with this model, but at this stage, we can note that Parsons thought that economics was a science of economising action with special reference to questions of adaptation between the environment and the social system. Parsons related adaptation to how systems function within an environment and conceives that the environment is affected and may be adapted to the society. The requirement to achieve the goal within the environment is ensuring enough resources exist to achieve a specific outcome. Therefore, adaptation occurs within the economic sphere within which adaptation should take place. For example to enable climate change through rural youth, adaptation could include obtaining work opportunities for them towards areas that enable them to sustain their environment. The nurturing of such an economic and social environment are important drivers to realise this goal. The economy has categories that enable its environment such as agriculture, indigenous innovation and services, whereas the social environment can relate to family, schools and community. The social and economic systems function to enable goal attainment. However, the influence of these systems and particularly relating to the climate justice agenda is more complex and requires in-depth exploration.

Undoubtedly, the climate disturbances that affect a system under consideration provide the reason for adaptation. Both the characteristics of the biophysical disturbance and of the affected system are relevant for adaptation. However, for the core concepts of the action theory of adaptation, it is not sufficient to consider social processes alone, as might be appropriate for a purely sociological issue. For investigating adaptation to climate change, we need to discuss the relation of actions beyond social processes and actions, since the interlinkages to the natural environment are crucial. We must deal with an interdisciplinary problem of interlinked biophysical and socio-economic systems. System characteristics determine adaptation processes that can result in adaptedness as outcome. They point out some key areas for research to improve adaptedness, in particular the identification of barriers to implementing adaptation and the governance of adaptation.

Just Adaptation and the Systems Approach

Socio-economic transformation in society is an imperative if we are to experience positive outcomes of a climate-smart society. However, to achieve sustainable outcomes and just adaptation requires integrating efforts through a systems theoretical lens. Systems thinking focuses on how the things being studied interact with the other constituents of the system. Bronfenbrenner is one of the most renowned theorists on systems theory with his Bio-Ecological Systems Theory (BEST). Systems theories such as BEST have been applicable for designing adaptation assessments. Bronfenbrenner's (1979, 1995) Ecological Systems Model (ESM) was firstly referred to as the Human Ecology Theory, and later, as a result of addressing human biology within the model, it became known as the BEST. This model positions human development and behavioural change within a person–environment perspective, suggesting that there are influencers within the various physical and social environments within which humans interact, which impact on human development.

Bronfenbrenner (1995) recognises five system layers, each holding several systems and subsystems important in understanding human development and change. These systems with their subsystems are layered and referred to as:

- the microsystem (context of individual's immediate environment of influence, i.e. family, school, church, neighbourhood/community)
- the mesosystem (context of relations between microsystems, i.e. family members, members within a congregation, school staff)
- the exosystem (context of relations between a microsystem and a system not directly involved with an individual)
- the macrosystem (context of the culture in which individuals live and interact as influenced by religion, economy, politics, education, more specifically, attitudes, ideologies or belief systems)
- the chronosystem (i.e. the way in which environmental effects develop over time and/or transitions affect individuals' growth and development).

According to Bronfenbrenner (1979), these systems help in understanding people's evolving conception of their environment, their relationship with their environment and their capacity to discover, sustain and/or alter environmental properties (Bronfenbrenner, 1979).

A comprehensive systems approach is essential for effective decision-making with regard to global sustainability, since industrial, social, and ecological systems are closely linked. Despite efforts to reduce effects of climate change, global resource consumption and exploitation of resources and anthropogenic activities continue to increase. There is an urgent need for a better understanding of the dynamic, adaptive behaviour of complex systems and their resilience in the face of disruptions, recognising that steady-state sustainability models are simplistic. In considering the different systems within the environment, the primary contribution of BEST is in accentuating the nature of reciprocal interaction between individuals and their environments. This interaction enhances understanding of human development through providing insight into the dynamics or interplay between process (P), persons (P), context (C) and time (T). Collectively, these are referred to as the PPCT Framework. As proximal processes, a dynamic interplay or reciprocal interaction is created between the features of an individual, the effect of social contexts and impact of developmental processes over time (Bronfenbrenner & Morris, 1998), which encourage change and development.

The foregoing means a comprehensive systems approach is essential for effective understanding of the dynamic, just adaptive behaviour of complex systems and their resilience:

...human beings are the primary agency of social change. the rates at which the population grows, its geographical distribution and the proportions in which it is divided between farms and cities, the racial and national stocks from which it comes, its age trends, sex ratios and marital condition – all of these help to determine the rapidity and the direction of past and future changes.

(Stoetzel, 2006, p. 19)

The BEST provides avenues through establishing the vulnerability of natural and human systems to climate change and influencing the capacity of these systems to adapt. This means that within existing layers of systems, we are able to understand critical influences for adaptative capacity in natural ecosystems to climate change and how climatic and non-climatic threats may limit the scope of just adaptations. This refers to the economic and political influences as much as the social and cultural fibres of the systems. This resonates with Bronfenbrenner's person-context (PC) relational attributes as mechanisms to reinforce the need for planned adaptation. Capon, Chambers, Nally, and Naiman (2013) suggest that such planned adaptation mechanisms must be determined by underpinning effective systems for gathering and interpreting information to inform vulnerability and risk assessments to know how, where and when to act. Secondly, they suggest that the systems approach to adaptation must be done at a small scale and in a landscape context with consideration for restoration given to the most vulnerable areas. Lastly, the BEST allows us to plan and consider ecological systems functions which stakeholders (either from political system, multimedia, political), goods and services can be involved within just adaptation. However, other researchers have observed that since its inception, climate justice has mainly focused on large-scale phenomena involving multiple-impact channels and complex interactions (Arndt et al., 2012). This chapter opts to take a focused approach.

Considering the facets presented within the theoretical underpinnings of just adaptation within a systems context, I opted to utilise the vulnerable context of African youth within rural communities to understand their potential adaptive capacity to bring about change within their predicament and climate change.

Youth Agency and Climate Change

Durham (2000) opines that the intrinsic nature of the youth in Africa describes youth as a social category, a social landscape for empowerment and agency. Therefore, youth by their nature have generally been instrumental in leading movements such as radical social movements towards upholding fundamental rights and justice. Climate change disruptions demand the same agency for youth to carry forward the same momentum and even more to be advocates for adaptation. This is because a people-centred approach to climate justice is critical. This entails ensuring representation, inclusion and protection of the rights of those most vulnerable to the effects of climate change. In addition, the new green learning agenda proposes an approach for an education system that develops and nurtures sustainable mindsets, as well as green skills in order to achieve transformation. Youth have been known to be nurtured towards achieving transformational goals. It is for this reason that youth were considered as critical stakeholders in advancing just adaptation and exploring possibilities.

Consequently, majority of youth in Africa and more so in South Africa reside in rural areas. These rural areas by their nature are considered to be vulnerable communities. These rural areas, to a varying degree, lack adequate infrastructure to meet the needs of its population (Alston & Kent, 2009; Mayende, 2010), which, in turn, affects their quality of life. In addition, the high rate of poverty means they are more susceptible to not only injustices but also suffer more from the effects of climate change. Poverty creates a culture of being vulnerable, powerless, isolated and physically and spiritually inadequate, which goes against principles of human development (Seeman, 1959). As Marx would infer, poverty perpetuates alienation (Seeman, 1959). Consequently, rural youth have been reported to be living in poverty and prone to socio-economic challenges such as unemployment crime, unplanned pregnancies, drug and alcohol abuse and the relentless HIV and AIDS epidemic (Wilkinson et al., 2017).

South Africa and the rest of the African continent's youth are currently at risk of losing their future which in some cases can be described as bleak. According to the World Bank, by the time many of the teenage climate activists of today are in their late 20s, climate change could force an additional 100 million people into extreme poverty. In addition, by 2050, the International Food Policy Research Institute estimates a 20% increase in malnourished children compared to what we would see without climate change (UNICEF, 2011) The question remains as to what possibilities exist for youth within the midst of a climate change crisis and how can integration of agendas be achieved to enable just adaptation.

Rural Youth in Mhlontlo Local Municipality: The Case Study

A central feature of the emerging strategy of the South African government on rural development is its almost exclusive focus on rural areas, which is premised on the assumption that it is, in these areas, envisaged that development must take place (Mayende, 2010). The Comprehensive Rural Development Plan of South Africa (Department or Rural Development and Land Reform, 2009) presents an economic model that shows how agrarian transformation is to contribute towards improvement of the lives and conditions of rural populations, specifically, rural youth as one of the foci. In South Africa, young people are most affected by various social problems, and at the centre is unemployment, which reached its peak in 2022 at 35.3%, according to Statistics South Africa. In this instance, while mitigation mechanisms can be adapted, development and employment opportunities are conceivable in the rural communities.

Within this context, this chapter explores the experiences and perceptions of youth in a rural community of Mhlontlo Local Municipality to understand the possibilities and hinderances for adaptive capacity to enhance climate-smart actions. The biophysical aspects of the municipality were of critical importance to unpack the relation between context and adaptive measures. Selection of this locality was also influenced by factors such as the rural nature of the area, demographic considerations related to youth and the availability of willing research partners that enabled data gathering. With reference to the total population, children and youth constitute 71.03% (birth to 34 years of age) of the population (Mhlontlo Local Municipality, 2017, p. 33). This implies that current and future youth need interventions enabling an environment to facilitate youth development and adaptation.

Research Methodology

The youth sample was purposively selected from a predetermined target population. The primary target group for the study was youth respondents who were beneficiaries of the National Rural Youth Service Corps (hereafter NARYSEC), a government programme towards transformation in the rural areas. The database of the youth was sourced through the key informant for skills development in NARYSEC. It was useful to use the database with a total of 67 registered youth in Mhlontlo to obtain contact details of the selected youth respondents. From the primary target population, 15 youth respondents were purposively selected, within the age group of 20–34 to participate. In addition and in alignment with the systems thinking family members and representatives of actors within economic institutions, schooling and religious groups were targeted as key informants. Data was collected through qualitative methods of inquiry, namely focus group discussions and key informant interviews to build on perceptions and experiences of youth that may enhance or hinder possibilities for adaptation and/or their adaptive capacity.

Lives and Livelihoods of Rural Youth in a Changed Climate

Youths who participated in focus group discussions acknowledged the hard work and contribution of their families who worked in subsistence farming projects supported by the municipality. However, when prompted on their involvement in the projects, majority of the youth respondents indicated that they were not actively involved. One youth respondent mentioned they would pursue agriculture only as a last resort for survival as they did have the knowledge on how to farm. Other youth respondents also revealed having knowledge and skills that included thatching of houses, building kraals and herding cattle, as practices they were exposed to while growing up. The Rural Development Researcher also mentioned that youth had indigenous knowledge that was passed from generation to generation. She specifically mentioned youth's ability to work on the land, their understanding of breeding livestock and having cultural competencies to treat ailments, even survival strategies. According to the researcher, youth have potential to be innovative, as they have knowledge that their urban counterparts do not possess.

As a rural community, there is interest in subsistence farming. However it was also apparent from the focus group respondents that there is need for continued support to grow subsistence farming projects. Family respondents also highlighted that Mhlontlo Local Municipality contributed through enabling access to land for them to plant as well as to irrigate using Tina and Tsitsa rivers. These family respondents also indicated that they had arable land and fenced areas, which was an advantage for utilising the land for not only crop production but also livestock farming.

Apart from viewing land as a resource for farming, the ward councillor mentioned that rural areas are spacious, and formal housing is more accessible compared to urban centres. Statistics South Africa (2016) has reported on the effect of rural–urban migration on the increase of informal settlements in the urban areas of the country. Research provides evidence pertaining prevalence of a high-level health risk, increased carbon emissions in informal settlements (Naicker et al., 2015); it is for this reason that rural communities such as Mhlontlo Local Municipality are advantaged.

Mhlontlo Local Municipality resembles what Brann-Barret (2011) refers to as a marginalised community characterised by poverty and a high youth unemployed rate and youth who experience the influence of alienation. The state of alienation is demonstrated by powerlessness and isolation due to the effects of poverty, high unemployment and limited economic opportunities. In addition, with a high rate of unemployed youth, researchers such as Cloete (2017) have written convincingly to suggest that the state of youth unemployment in South Africa was perpetuating alienation. This is illustrated in a statement based on research that

> young people are beginning to feel a sense of alienation from the larger society and a sense of betrayal by the government, because they realise that their lives have not changed for the better since 1994.
>
> (Cloete, 2017, p. 514)

Seeman (1959) refers to alienation as a crisis of personal identity produced by inconsistencies between socio-personal microsystems, culture and structure of society itself. Moreover, he identifies with alternative meanings of alienation that include powerlessness and isolation, which are evident in Mhlontlo Local Municipality. In addition, based on the work of Granovetter (1983), fragmentation relates to weak ties that exist in social networks. Parkins and Angell (2011) go further to explain that fragmentation is identified with isolation and lack of broad networks of social support. Consistent with this definition, findings in Mhlontlo Local Municipality provide evidence of the weak ties in social connections and relationships which has contributed to unemployment and the social problems in the community. For example, in Mhlontlo Local Municipality, it was found that youth community-based programmes and other support systems for youth were not connected and often a duplication. This meant that the programmes had the same aim of developing youth and often repetitive in who they offered the opportunities to. Accountability processes are limited as they not only lack monitoring and evaluation but also relevant integrative processes.

The municipal projects in farming and other various agricultural activities were not associated with the skills development initiatives in the community-based programmes such as NARYSEC. Considering that land is a key resource in the community, there is a need to optimise youth participation in resourceful activities such as farming through involving community stakeholders. In addition, policy documents such as the Comprehensive Rural Development Plan (2009) support processes of agrarian transformation in rural areas, proving the committed efforts of government to develop rural communities. However, while Mayende (2010) does well in his research to illustrate the limitations in the implementation of the Comprehensive Rural Development Plan, he does not also address the need for integrative processes to support its implementation through community participation.

Apart from agricultural productivity as potential investment, no mention of recycling or climate-smart actions have been incorporated into local policies or projects. The developmental aspects are looked at in isolation of the pressing global questions of climate initiatives and responses to insight adaptative capacity. Nonetheless, the extent to which youth face alienation and exclusion was evidently a hindrance in considering rural youth adaptive capacity which highlighted a systems failure within the context for sustainable just adaptation.

Implications for Climate Justice and Adaptation

As indicated above, the life experiences of youths indicate several nuances that have implications for youth development and which impacts on just adaptation. The physical environment such as the impact of climate change impacts youth potential to develop unless their adaptive capacity is enhanced. The state of social alienation of rural youth if not addressed has implications for their self-development and measures of climate justice within these areas wherein the effects are severely felt. The implications for development and climate adaptation require a system thinking as proposed below based on the findings.

Connections within and between Networks for Youth

Youth have considerable associations and connections that influence their day-to-day activities; however, the networks are often weak. Youth need more exposure to life beyond their experience of their home and school; this enables youth to acquire a well-encompassed enabling environment. This would broaden their views and learn lessons from such experiences.

In developing adaptive and development avenues, youth must be enabled to enhance positive character strengths, competence socially and academically, caring for others and confidence in themselves. This is enabled by an environment that supports and nurtures its young people. Creating an environment for youth to develop positively and thereby become empowered requires both youth and the community to engage in participative processes that provide support. Integration of agendas, such as climate justice and youth empowerment, is proposed as a process to support youth through merging together different initiatives such as strengthening agricultural activities in rural communities.

In most of Africa, and more so South Africa, most rural areas are dependent on small-scale subsistence farming (Bennell, 2007; Lohnert & Steinbrink, 2005). However, there is debate as to the extent to which rural communities have continued to be active in subsistence farming, particularly the involvement of youth in South Africa. Farming is considered a large untapped reservoir of opportunities, which rural communities have an advantage over, due to the availability of space and fertile lands (World Bank, 2014). Mayende (2010) also suggested that rural areas have the capacity to forego agrarian transformation to eradicate the high rate of rural poverty that characterises most rural communities in South Africa. Agrarian studies have for decades sought to understand the diverse paths of transition from non-capitalist peasant economy towards capitalism in the Global South; particularly, in the wake of post-1980s neoliberal restructuring, this complexity is still a conundrum in South Africa to be explored further.

Collective Ownership and Responsibility

On the other hand, the research respondents in Mhlontlo Local Municipality emphasised that for youth to develop positively required a change in the attitude of the youth towards self as well as social responsiveness. This also suggests the need for collective responsibility to ensure youth develop positively and enhance their adaptive capacity. Holzner (1967) suggests integration for effective collaboration of not only social actors but also physical resources in the environment and institutions. Bronfenbrenner (1979) provides the most suitable exposé of the ecological systems and what integrative aspects may be considered when creating an enabling adaptive environment.

Using an integrative imaginative approach depicts how to integrate resources (human, services or materials) within a system that recognises the impact of all facets of the environment. The state of social alienation that the youth face is a result of a fragmented system that does not tie the linkages between the physical locality dynamics to the socio-ecological dynamics. An integration of the physical and ecological manifests a biophysical and psycho-social dynamics to inform strategic interventions that can impact adaptive and development outcomes.

Therefore, youth in rural areas have adaptive capacity that exists through the strong networks that already exist within the system, and these need to be harnessed towards just adaptive projects in order for youth to benefit in both development and sustainability initiatives. These initiatives must be weaved into existing systems to provide for human plasticity and agency for positive outcomes in the vulnerable communities and groups.

Conclusion

This chapter provided insights on the actions of youths in response to climate impacts and attendant climate justice issues. In doing so, it is evident that weak ties exist between local policies in terms of addressing climate justice as there are weak ties in understanding the contextual variability of the nature of the vulnerable context. On this, it also shows that vulnerability exists in three contexts, namely a group of individuals with a similar bio-social dynamic and the environmental vulnerability of geographical location and lastly that of accessibility. Climate justice practitioners should invest in understanding these variabilities that exist in context if all efforts towards climate justice are to succeed.

References

Alston, M., & Kent, J. (2009). Generation X-pandable: The Social Exclusion of rural and remote young people. *Journal of Sociology, 45*, 89–107.

Arndt, C., Chinowsky, P., Robinson, S., Strzepek, K., Tarp, F., & Thurlow, J. (2012). Economic development under climate change. *Review of Development Economics, 16*(3), 369–377.

Bennell, P. (2007). *Promoting livelihood opportunities for rural youth.* International Fund for Agricultural Development: Rome.

Brann-Barrett, M. (2011). Same landscape, different lens: Variations in young people's socio-economic experiences and perceptions in their disadvantaged working-class community. *Journal of Youth Studies, 14*(3), 261–278.

Brisley, R. W. (2012). *An exploration of how far social justice is considered in local adaptations to climate change impacts across the UK.* London: Joseph Rowntree Foundation.

Bronfenbrenner, U. (1979). *The ecology of human development: Experiements by nature design.* Cambridge: Harvard University Press.

Bronfenbrenner, U. (1995). The bioecological model from a life course perspective: Reflections of a participant observer. In P. Moen, G. Elder, & K. Luscher (Eds.), *Examining lives in context: Perspectives on the ecology of human development.* Washington, DC: American Psychological Association.

Bronfenbrenner, U., & Morris, P. (1998). The Ecology of Developmental Processes. In W. Damon, & M.L. Richard (Eds.), *Handbook of child psychology: Theoretical models of human development* (pp. 993–1023). New York: John Wiley and Sons, Inc.

Capon, S., Chambers, L., Nally, R., & Naiman, R. (2013). Riparian ecosystems in the 21st century: Hotspots for climate change adaptation. *Ecosystems, 16*, 359–381.

Cloete, A. (2017). Youth unemployment in South Africa, A theological reflection through the lens of human dignity. *Missionalia Journals, 43*(3), 513–525.

Department of Forestry Fisheries and Environment. (2018). *South Africa's low emission development strategy 20250.* Pretoria: National Government of South Africa.

Department or Rural Development and Land Reform. (2009). *Comprehensive rural development plan.* Pretoria: The South African Government.

Durham, D. (2000). Youth and the social imagination in Africa: Introduction to parts 1 and 2. *Anthropological Quarterly, 73*(3), 113–120.

Eisenack, K., & Stecker, R. (2012). *An action theory of adaptation to climate change.* Berlin Conference on Human Dimensions of Global Environmental, Berlin.

Granovetter, M. (1983). The strength of weak ties: A network theory revisited. *Sociological Theory, 1*, 201–233.

Holzner, B. (1967). The concept "integration" in sociological theory. *The Sociological Quarterly, 8*, 51–62. Retrieved November 17, 2017, from http://www.tandfonline.com/doi/abs/10.1111/j.1533-8525.1967.tb02273.x?journalCode=utsq20.

Intergovernmental Panel on Climate Change. (2001). *Climate change 2001: The scientific basis.* Cambridge: Intergovernmental Panel on Climate Change.

King, N. (2021). *Climate change implications for SA's youth.* Pretoria: Centre for Environmental Rights. Retrieved November 4, 2022, from www.cer.org.za

Lohnert, B., & Steinbrink, M. (2005). Rural and urban livelihoods: A translocal perspective in South Africa context. *The South African Geographical Journal, 87*(2), 95–103.

Mayende, G. (2010). Transforming labour reserves in South Africa: Asymmetries in the new agrarian policy. *African Sociological Review, 15*(1), 49–71.

Mhlontlo Local Municipality. (2017). *Mhlontlo local municipality integrated development plan 2017–2022.* Review. Provincial Government of the Eastern Cape South Africa.

Naicker, N., Mathee, A., & Teare, J. (2015). Food insecurity in households in informal settlements in urban South Africa. *The South African Medical Journal, 105*(4), 268–270.

Parkins, J., & Angell, A. (2011). Linking social structure, fragmentation, and substance abuse in a resource-based community. *Community, Work and Family, 14*(1), 39–55.

Rawls, J. (1971). *A theory of justice.* Oxford: Oxford University Press.

Schlosberg, D. (2012). Climate justice and capabilities: A framework for adaptation policy. *Ethics & International Affairs, 26*(4), 445–461.

Scholes, R., & Engelbrecht, F. (2021). *Climate impacts in southern Africa during the 21st century.* Johannesburg: Global Change Institute.

See, J., & Wilmsen, B. (2020). Just adaptation? Generating new vulnerabilities and shaping adaptive capacities through the politics of climate-related resettlement in a Philippine coastal city. *Global Environmental Change, 65*(5), 1–13.

Seeman, M. (1959). On the meaning of alienation. *American Sociological Review, 24*(6), 783–791.

Statistics South Africa. (2016, April 20). *GHS series volume VII: Housing from a human settlement perspective.* Retrieved from StatsSA: http://www.statssa.gov.za/?p=6429.

Stoetzel, J. (2006). Sociology and demography. *CAIRN.INFO International Edition: Population, 61*(1), 19–28.

Trading Economics. (2023). *South Africa GDP growth rate.* Retrieved January 22, 2023, from https://tradingeconomics.com/south-africa/gdp-growth

UNICEF. (2011). *Exploring the impact of climate change on children in South.* Pretoria: UNICEF South Africa.

United Nations. (2015, November 4). *Paris Agreement.* Retrieved from United Nations *Climate Change*: https://unfccc.int/process-and-meetings/the-paris-agreement

United Nations. (2020). *United Nations climate change annual report 2020.* Retrieved January 20, 2023, from https://unfccc.int/sites/default/files/resource/UNFCCC_Annual_Report_2020.pdfUniversity of Cape Town. (2017). South Africa and the G20: Where Do We Stand on Greenhouse Gas Emissions? Retrieved January 27, 2023, from https://www.news.uct.ac.za/article/-2017-07-05-south-africa-and-the-g20-where-do-we-stand-on-greenhouse-gas-emissions

Wilkinson, A., Pettifora, A., Rosenberga, M., Halperna, C., Thirumurthya, H., Collinsonf, M., & Kahnf, K. (2017). The employment environment for youth in rural South Africa: A mixed-methods study. *Development in South Africa, 34*(1), 17–32.

World Bank. (2014). *Youth employment in sub-Saharan Africa.* Washington, DC: World Bank.

11 Climate Change and Justice in Botswana's National Adaptation Strategy and Programming

The Journey So Far

Stanley O. Ehiane and Christopher Dick-Sagoe

Introduction

There is a growing debate at the national and international levels on climate change and its attending impacts on livelihood security, migration and displacement, and agriculture and conflict. The growing global warming and the greenhouse emission from humanity culminate in climate change and environmental degradation (Adedeji, Okocha, and Olatoye, 2014). The Intergovernmental Panel on Climate Change (IPCC) has asserted the world climate influenced by human activities in the last century (IPCC, 2007). However, some of the impacts of climate change include drought, variation in the precipitation pattern, rise in the sea level, land degradation, and water shortage, which are all presently being felt but to be worsened in future (Skah and Lyammouri, 2020). The climate is a long-term weather pattern that describes a particular region. But the word climate change connotes a change in global temperature resulting from "natural climates like the large eruption of the volcano and sun's energy and persistent greenhouse gases and emission of black carbon to the atmosphere" (Adedeji, Okocha, and Olatoye, 2014: 2). Meanwhile, the variability in climate across the world occur due to natural internal processing and/or human driving forces (Anthropogenic) like the burning of fossil fuel within the climate system (Adedeji, Okocha, and Olatoye, 2014), and the climate variability could persist for many years.

Climate variability accounted for the extent of the impact of climate change on livelihood as it determines how often a region experiences a rise in sea level, flood, drought, and heatwave and it varies across the world (Behrens, Georgiev, and Carraro, 2010). This variation in the global average surface temperature is what is referred to as global warming also global cooling. For quite sometimes the earth has been witnessing changes in the intense heat of the sun causing the cycle of warming and cooling. Though climate change occurs naturally increased human-induced greenhouse gases in turn increase the warmness of the natural greenhouse effect potentially increasing the rate of global warmness that was never experienced by mankind, particularly in Africa (Adedeji, Okocha, and Olatoye, 2014). Nevertheless, the concept of climate change and global warming are often used interchangeably, the word climate change is more appropriate as it encompasses human and natural phenomena (Frimpong, 2020).

DOI: 10.4324/9781003397120-13

The consequences of climate variability varied from one region to another across the globe. Also, it is reflected in the individual region's access to adaptation mechanisms, finance, technology, and climate information and services (Skah and Lyammouri, 2020). Climate change has been considered the leading human and environmental problem in recent times, particularly in Africa as currently witnessed. Of course, scholars, politicians, and diplomats have established that climate change poses substantial threats to the continent including security threats (Ki-moon, 2009). This assertion could be attributed to weak awareness, finance, technology, and structural factors (Behrens, Georgiev, and Carraro, 2010). Africa has suffered disproportionately in the pursuit of food, water, sustainable development, and socioeconomic sustainability due to climate change, despite the continent being the least of global pollutants (Tadesse, 2010). However, adaptation and mitigation options are a matter of sustainability as the impacts of climate variability already happening. Deforestation, desertification, high temperature, drying up of soil, flood, and erosion in Africa are the impacts of climate on agriculture (Tadesse, 2010).

The impacts on agriculture have huge consequences on the economy and the living nature of the people in Africa (Adedeji, Okocha, and Olatoye, 2014), because of the overreliance on agriculture. The economic impacts of climate change inducement besides the fact it is substantial, its multiplier effects cannot be overemphasized particularly in the industries that solely relied on primary produces for raw materials (Stern, 2006). Despite scholars being unable to quantify the economic cost of climate change in the individual country and the continent. The economic cost of living according to scholars has been predicted to "increase from 5 per cent to 20 per cent of the world income" (Adedeji, Okocha, and Olatoye, 2014: 5). However, to address the impacts of climate change in terms of mitigation and adaptation on the countries in the South, the need for climate justice is significant. The emergence of climate justice is a response to the need for a novel approach to climate change justice, particularly challenges. In light of the aforementioned, this study explores climate change and its impact on Botswana, further giving a snapshot into Botswana's climate change policy framework. The study also unpacks climate justice capability, the framework for assessing climate justice and Botswana's adaptation strategy. The paper anchored its analysis using the climate justice framework to assess the Botswana climate change strategy and examines the challenges inherent in achieving climate justice in Botswana.

Climate Change Impacts in Botswana

It is evident that climate change affects every part of the planet, although Africa and Asia are significantly more vulnerable (UNICEF, 2018). Now, viral and vector-borne diseases such as dengue, hepatitis, and malaria are rapidly spreading. The United Nations International Children's Emergency Fund (UNICEF, 2018) estimates that by 2050, 600 million people in Africa would be water-stressed. If climate change is not mitigated, it will lead to significant security issues, just as it did in Darfur, a prime example, according to Joshua (2021), where a conflict began as an ecological crisis, partially fuelled by climate change and a 20-year Sahelian drought. Joshua (2021) concurs that climate change necessitates a worldwide response and should be considered a danger to national security and human rights. Chaudhary and Mooers (2018) says that the uneven impact of climate change across the globe is one reason for the focus on the connection between climate change and human rights. Natural sensitive or vulnerable developing nations, such as Botswana, are already witnessing worsening weather patterns and far-reaching effects of climate change that hinder economic development and exacerbate the already prevalent poor situation. Food prices will continue to grow due to climate change. It will exacerbate interstate and intrastate strife, especially in regions where political instability and socioeconomic rifts were already common, and it will eventually spread across continents and even to formerly calm nations like Botswana. There are two potential ways in which the connection between conflict and the environment can affect security. Alternatively, disagreement may degrade the resources and services that the ecosystem provides, hence reducing security.

According to the United Nations Development Programme (UNDP, 2014b) Botswana's key concerns include increased energy and water stress due to rising temperatures and shifting precipitation patterns, losses in rangeland productivity and lower agricultural yields, which constitute a grave threat to food security. The temperature is anticipated to climb between 1°C and 3°C by 2050, resulting in higher potential evaporation rates (Boko *et al.*, 2007). Future patterns in precipitation are uncertain, however, the vast majority of general circulation models forecast a decrease in precipitation, maybe accompanied by more powerful showers locally. Botswana is extremely concerned about desertification, and the IPCC predicts that the share of arid and semi-arid countries in Africa would increase by 8% by the 2080s (Boko *et al.*, 2007). Water scarcity, water stress, and land degradation will have negative effects on Gross Domestic Product (GDP), poverty, health, and agricultural output (Government of Botswana, 2001). Botswana's ecosystems, particularly the Okavango Delta, are anticipated to be impacted by climate change, which is likely to have a detrimental influence on tourism and livelihood prospects for the inhabitants of the basin. In the coming years and decades, climate change consequences are predicted to intensify, posing a threat to development and diminishing the likelihood of meeting the Millennium Development Goals (MDGs) (Boko *et al.*, 2007).

Botswana's climate is arid to semi-arid, with warm winters and blistering summers with highly irregular precipitation. The majority of the nation's precipitation occurs between October and April (Government of Botswana, 2001). According to the 2020 University of Notre Dame Global Adaptation Initiative (ND-GAIN) Index, Botswana is ranked 94 out of 181 countries in terms of climate change vulnerability. This vulnerability results from a combination of political, geographical, and social factors (World Bank Group, 2021). University of Gothenburg (2008) indicates that Botswana has both national and transboundary conflicts over natural resources. Local conflicts over land; concerns from international experts and local populations over the ecology of the Okavango Delta in Botswana; Namibian plans to construct a hydroelectric dam at Popavalle (Popa Falls) along the

Angola–Namibia border; Botswana has built electric fences to stop the thousands of Zimbabweans fleeing to find work and escape political persecution and economic disintegration; Namibia has long supported, and in 2004 Zimbabwe dropped objections to, the construction of a pipeline from Angola to South Africa (Boko *et al.*, 2007). To avert conflict over transboundary resources, formal agreements have been made with the goals of prudent usage, environmentally sound development, and equal access to water resources.

The world's undernourished population will increase exponentially, placing strain on the limited resources, and this will eventually aggravate interminable worldwide conflicts. The primary sectors requiring policy interventions in adaptation are water resources, agriculture, livestock, health, forestry, biodiversity, disaster preparedness, vulnerable ecosystems, and socioeconomic measures. A modest increase in Botswana's average temperature as a result of climate change might have severe effects on human, plant, and animal life. Forecasts of the IPCC (2018) indicate that in Botswana, the consequences of rising global and local temperatures would be seen in various sectors crucial to the development of the people and economy. Understanding when different warming levels will occur and what these mean for threats to vulnerable sectors like agriculture, health, and water is vital for adaptation planning and identifying what must be done by when (Nkemelang, New, and Zaroug, 2018). According to the World Health Organisation (WHO) Botswana is warming faster than the rest of the world and rapid action is required (WHO, 2015). According to climate estimates, Botswana's temperatures are rising faster than those of the majority of countries around the world. Without concerted action to adapt to a hotter, drier future, Botswana's people, ecosystem, and economy all suffer (MEWT, 2012). In response to climate change, communities and the government must implement urgent and long-term measures to boost resilience and well-being.

Future projections indicate that Botswana will become drier, water-stressed, and more prone to climate extremes, including more frequent and violent droughts, floods, and heatwaves (Nkemelang, New, and Zaroug, 2018). Botswana could be 2°C hotter than pre-industrial temperatures by 2024, which is a considerably more rapid increase in temperature than other regions of the world. Increasing temperatures will affect numerous sectors, including agriculture, health, and water, which are essential to the economic and social well-being of humans (Adaptation at Scale in Semi-Arid Regions, 2018). Agriculture is particularly vulnerable to climate change, and farmers are likely to experience diminished crop yields and increased animal losses. The principal source of income in the Bobirwa subdistrict, rain-fed agriculture, is already marginal, and even if global warming is limited to 1.5°C Cover pre-industrial levels, forecasted climatic change may render current agricultural practices impossible (Nkemelang, New, and Zaroug, 2018).

The growing severity of Botswana's climate consequences from 1.5°C to 3°C requires concerted local and international action. Botswana must anticipate and prepare for the region's rapid weather and climatic changes. The development of adaptation solutions must be sped up, and they must be implemented in a way that benefits all individuals and economic sectors. The time for pilot adaptation efforts and experimentation has passed, and the time has come to incorporate climate resilience throughout the public, corporate, and community sectors. The Paris Agreement aims to keep global warming far below 2°C, and ideally below 1.5°C. Understanding the local-level consequences of these global temperature targets is vital for determining adaptation requirements and activities to climate change (ASSAR, 2018). The world is currently on course to warm by 3.2°C by the end of the century although states have made inadequate mitigation efforts. Botswana will see a local warming and drying that exceeds the world norm. Therefore, even a 1.5°C increase in global temperature will have severe negative effects on water supply, agriculture, public health, and other susceptible sectors. The 1.5°C threshold might be crossed during the next decade, and the 2°C threshold could be crossed the decade after that (Nkemelang, New, and Zaroug, 2018). Consequently, Botswana's adaptation measures must be expedited immediately (Kgathi, 2018).

On September 25, 2015, the 193 member states of the United Nations General Assembly, including Botswana, endorsed the "Transforming our world: the 2030 Agenda for Sustainable Development." 17 Sustainable Development Goals (SDGs), 169 linked goals, and 232 indicators comprise the Global Agenda. The 2030 Agenda and its 17 SDGs are based on the MDGs, but have a considerably broader scope, encompassing the Social, Economic, and Environmental components, whereas the MDGs focused primarily on the Social and Economic elements (UN, 2015). The United Nations and its partners in Botswana are working to achieve the SDGs, a set of 17 interrelated and ambitious objectives that address the most pressing development issues facing people in Botswana and around the world. Based on the agreement known as the United Nations Sustainable Development Framework (UNSDF) for 2017–2021, the collective aspirations of the United Nations in Botswana are to move towards greater collaboration, focus, and coherence in programming, and to improve the quality of life for all Botswanans, especially the most vulnerable groups. In support of inclusive, equitable, and sustainable development in Botswana, the Government and the United Nations (UN) in Botswana promise to work closely together to support the implementation of Vision 2036, National Development Plan (NDP) 11, and the 2030 Agenda for Sustainable Development at the national level.

Botswana's Climate Change Policy Framework

Botswana has produced a Climate Change Policy and Institutional Framework, which is accompanied by a Strategy and Action Plan to implement the Policy (approved in 2016). Botswana intends to ensure that the Policy can be implemented by developing a comprehensive package of measures that includes a long-term low-carbon strategy, a national adaptation plan, nationally appropriate mitigation actions, identification of key technologies, a plan for knowledge management capacity development, education and public awareness, and a financial mechanism (Ministry of Environment, Natural Resources Conservation and Tourism, 2020). Botswana has reached numerous climate change response milestones. It is a signatory to the United Nations Framework Convention on Climate Change (UNFCCC), the multilateral instrument that enshrines the resolve of the international community to combat climate change. Botswana is a signatory to both the Kyoto Protocol, which is about to expire and the Paris Agreement, which is currently in effect (UNFCCC, 2015). Botswana has filed two National Communications and its nationally determined contributions (NDC) as part of its UNFCCC commitments (UNFCCC, 2015).

Additionally, Botswana has created two national GREENHOUSE GAS (GHG) inventories. Currently, it is drafting its Third National Communication (TNC). In Botswana, climate change has received increasingly more attention from the government, especially as Climate change remains a challenge to the government's poverty alleviation drive. The eleventh NDP for the period 2017–2023 identified climate change as a challenge faced during the tenth NDP and emphasized the need to address climate change. Under the core focus area of "Sustainable Use of Natural Resources," the NDP stressed the importance that climate change mitigation and adaptation may play in job creation and economic growth, as well as the necessity of integrating climate change into development planning. In addition, the NDP called for gender-sensitive adaptation and mitigation measures and intelligent agriculture (the Republic of Botswana, 2019). In 2019, the Ministry of Environment, Natural Resources Conservation, and Tourism (MENT) established a climate change coordinating unit to assist in the management of several national and international climate change initiatives. The Department of Meteorological Services (DMS) was selected as the country's focal point for climate change response and to assist in meeting UNFCCC responsibilities. In addition to these policies, legal, and institutional developments, Botswana's establishment of a National Committee on Climate Change (NCCC) stands out as a significant step in its response to climate change. The NCCC is meant to serve as a consultative group to aid the DMS in executing climate change-related activities, especially on a technical level. This type of national body devoted to climate change is a significant acknowledgement of the need to focus on the issue and adopt a multidisciplinary approach to the problem.

The 2018 National Climate Change Strategy supports Botswana's Vision of becoming a sustainable, climate-resilient society whose development follows a low-carbon development pathway in pursuit of prosperity for all. The vision of the strategy is to provide the necessary guidance for Botswana to achieve a low-carbon footprint, as well as climate change awareness and resilience. Given that the strategy's intended timeframe is from 2018 to 2030, priority strategies are identified for each major sector in Botswana and then broken down into four sequential target actions, one for each of the years 2020, 2023, 2026, and 2030. Eleven sectors are prioritized for adaptation responses, while seven are prioritized for mitigation responses. Agriculture; water; human health; human settlement; forest, savannah, and woodlot; land use; disaster risk; biodiversity and ecosystem; infrastructure; manufacturing; and tourism are the sectors identified in the adaptation response. Sustainable energy; transportation; waste management; agriculture, forestry, and land use; extractive and mining; carbon budgets and abatement pathways; and market-based mechanism. In addition, strategies are identified for the following transversal themes: gender, education and training, equality and equity, innovation, research and development, communication and knowledge management, climate services, institutional arrangement, and resource mobilization.

Unpacking Climate Justice Capability

The debate on the cause and effect of climate change is on the increase at both global and local levels. The reality of the effect of climate change and adaptation ushered a major discourse that dominates climate change policy known as climate justice. The emergence of the varied understanding of climate justice in recent years has impacted the strategy adopted by governments on how climate justice is being implemented (Schlosberg, 2012). Justice in climate change considers who gets the benefits and who bears the cost of climate action and inaction. However, cause and effects, together with the strategies to address them through adaptation and mitigation boil down to climate justice and fairness (Homberg and McQuistan, 2019). Vulnerability to the impact of climate change is felt disproportionately by people with systematic marginalization and exclusion (Levy and Patz, 2015), whereas those considered most exposed to the impact of climate change are the least-developed countries of the global South (Homberg and McQuistan, 2019). This has been considered true due to myriad factors such as economic and reliance on climate-threatened economic activities such as livestock and farming in most developing countries (Homberg and McQuistan, 2019).

Of course, "there are ways for harming others by action without intention and are geographically remote," this recognizes disproportionate impacts of climate change on different groups (Schlosberg and Collins, 2014: 28). Again, some countries are more vulnerable than others to the impact of climate change, but vulnerability is not limited to injustice but the capability to respond to the impact of climate change. Such capability includes socioeconomic, environmental and cultural factors, and institutional practices such as planning rules and housing policy (JRF, 2014). The global injustice regarding the causes and effects of climate change led to the calling for differentiated responsibility between the 'North and South' in addressing climate change globally (Schlosberg and Collins, 2014). Therefore, climate justice is an avenue to strike political consensus considered to be a fair result in sustained action by most countries, being a transformative change beyond national boundaries. That said, decision-making at both global and national levels towards the culture of climate justice must be open, transparent, and corruption-free (MRF, 2019).

At the local and national level, climate justice has been one political dialogue in the design of climate change policies. The need to unpack the relationship between the climate justice concept and the effects climate change has on vulnerable communities have been the focus of climate theorists, activists, and government of developing countries (Baxi, 2016). The popular models which have been developed based on the relationship explained are the polluter pay, fair share, and the right-based (Baxi, 2016). Polluter pay models fundamentally draw from historical responsibility and opine that the party responsible for the current climate change crisis should pay more for their transgression (Schlosberg, 2012). This translate that least-developed countries are more affected by the impact of climate change than developed countries. Hence, developed nations responsible for climate change should play a greater role in preventing and mitigating its impact on vulnerable groups (Schlosberg, 2012). The fair share model argues for the equal allocation of emissions. It proposes the total amount of greenhouse gas to be emitted divided by the global population, and each country will be allowed to emit the sum of its population. Therefore, the country with high emissions will be allowed to buy from those with low emissions, and this serves as compensation to the nation with lower emissions.

Framework for Assessing Climate Justice

As identified by the United Nations, climate change's impacts are not borne equally and fairly, as stratifications such as rich and poor, men and women, and older and younger generations exist in society. Environmental justice is the mother theory from which climate justice finds its root. Climate justice has both substantive and procedural components (Dolšak and Prakash, 2022). Procedural concerns itself with the inclusive active participation of the public in climate justice. This then makes provision for access to solutions which seek to correct the harm citizens receive as a result of policies put in place to tackle climate change (Schlosberg and Collins, 2014). Substantive on the other hand considers the observable inequalities in cost and benefit distributions from climate inaction and action.

Three dimensions are offered here as a framework for assessing climate justice from climate action and inaction perspectives. Out of the three dimensions, the first touches on climate inactions. Climate change disproportionately affects different people and communities, exposing these underprivileged communities to the impact of climate change. The impact of climate change manifests in extreme weather events, prolonged drought, and rising sea levels. Wealthy countries benefitted from fossil fuels to power their industries for production (industrialization), thus contributing more to climate change. Yet the impact of climate change is more felt in less developed countries (less wealthy) and hardly contributed to climate change dues to less use of fossil fuel to power industries, whereas the main contributors (developed and wealthy nations), some even benefit from the impact of climate change. Besides, others have enough financial resources to acquire technology which can reduce and absorb the impacts of climate change.

The rich–poor split is also seen in wealthy nations (and even within cities). In comparison to the poor, wealthy households who have profited from the industrial economy are more protected from the effects of climate change. Furthermore, gender plays a role in climate change in disadvantaged places, for which women in developing nations face additional challenges due to climate change that have an impact on their economic, social, and physical well-being.

The second-dimension touches on the unequal distribution of costs associated with the government's climate change mitigation and adaptation policies. Governments develop policies to mitigate and help their people to adapt, especially those vulnerable to the impact of climate change, to deal with the effects associated with climate change. These policies differ in how they allocate benefits and cost across various communities and sectors, leading to distributional disputes that frequently reinforce existing disparities (Dolšak and Prakash, 2022). As a result, the second aspect of climate justice concentrates on the unequal distribution of costs associated with mitigation and adaptation strategies. Measures taken to regulate energy, transport, and the use of fossil fuels are necessary for decarbonization. However, such measures as carbon prices are not financially favourable to the poor because they spend a large chunk of their income on energy.

Pollution may be directed towards underprivileged areas by cap-and-trade legislation. Global emission trading may cause emissions to move from developed to developing nations. Mandates for renewable energy sources increase mining in developing nations, which is costly for the environment, culture, and human health (Sovacool *et al.*, 2020). Through adaptation, governments sometimes disproportionately condemn the homes of low-income families when they pursue managed withdrawal from flood-prone locations. Building regulations that increase resilience becomes costly for low-income people to afford.

Lastly, the third dimension focuses on the unequal distribution of the benefits of climate policies. Decarbonization is set to lay the groundwork for a new economy; however, it is uncertain who gains from this new economy. The reason is that skill sets are not transferable and retooling the workforce is mostly difficult. Is it possible that today's coal miner will be tomorrow's solar panel installer or turbine technician? This is not possible. In addition, there are social and financial implications for moving from one work in one sector to another job, probably in another sector. Further, wages in emerging industries might not be comparable, particularly if union rates are low.

New technologies are introduced by decarbonization, yet the affordability of such technology to low-income people will be another challenge. Compared to cars with internal combustion engines, electric vehicles are pricy. The process of recharging electric vehicles exacerbates equity issues. Electric vehicles are practical for people who can install chargers at home because recharging takes a while (Guo and Kontou, 2021). Additionally, rooftop solar favours residents of single-family dwellings. The wealthy and privileged may benefit from government support for recovery from natural disasters.

Climate injustice, according to Concern Worldwide (2022), manifests in ten major areas and warrants the attention of the National Climate Change Adaptation Policy, including that of the Government of Botswana. These injustices are geographic injustice, settler colonialism and indigenous exploitation, gender inequalities, intergenerational climate injustice, economic inequalities, racism, language and literacy barriers and immigration status, discrimination against the disabled, chronically ill, and elderly, and other forms of societal inequality and inequality of resources beyond climate mitigation.

The explanation for these points begins with geographical injustice. At the local level (country-specific Botswana), most of the human activities which contribute to climate change are located in the urban areas and are controlled by the rich, whereas activities such as farming, mostly located in rural areas, are usually owned by the poor subsistence farmers. This poor farming continues to fall prey to climate change impact. On gender inequalities, it was realized that women face the brunt of climate disasters and therefore any effort to consider climate adaptation policy should focus on women. Youth (intergenerational climate injustice) are to be considered in climate change adaptation with the argument that the youth will live longer with the impact of climate change compared with the elderly. On economic inequality, the poor usually live on agriculture or pastoralism which is a target activity for climate disasters. Above all 70% of the global estimate for extremely poor people are women (World Bank, 2021). Elderly, chronically ill people and disabled people face a high risk of death from climate-related disasters and therefore should be factored in the planning for climate change adaptation, thus ageism contributes to increased climate change vulnerability.

Botswana's National Climate Change Adaptation Strategy

The country maintains a commitment to developing a long-term strategy that emphasizes low-carbon emissions, support for the mitigation of climate change and adaptation strategies (World Bank Group, 2021). The government 2009 developed the National Disaster Risk Plan. The plan focused on prevention, mitigation, preparedness, response, and recovery strategies. The plan aimed to enable sustainable development by carrying out disaster management strategies, reducing vulnerability, and increasing resilience within communities to climate change (World Bank Group, 2020). In 2014, the government submitted the Botswana Climate Change Response draft to the UNFCCC. The draft expands on the framework of the disaster risk management plan and further articulates the government's commitment and the strategies it proposes to reduce Botswana's vulnerability to climate change as well as protect the livelihoods of the population (UNDP, 2014a).

The strategic vision of the Botswana National Climate Change Final Strategy submitted to the UNDP in 2014b is "a society that is sustainable climate resilient and whose development follows a low carbon development pathway in pursuit of prosperity" (UNDP, 2014b). The key components of this strategy include sustainability of the environment, water resources, sustainable land management, agriculture, and health (World Bank Group, 2020). The strategy considers the effects of climate change on gender and ensures that policies are gender sensitive by recognizing women's and youths' vulnerability to climate change (UNDP, 2014b). The adaptation strategy focuses on sectoral policies and legislation in the sectors of agriculture and food security, water, forestry, human health, land use, human settlement, disaster risk management, biodiversity and ecosystems, infrastructure development, industry and manufacturing, and tourism which are identified as the most affected by climate change and heavily impact human security (UNDP, 2014b). The final strategy addresses the key sectoral vulnerabilities and posits the strategies to overcome these threats as well as climate mitigation strategies across all sectors.

Botswana also relies on regional and continental strategies such as the Africa Regional Strategy for Disaster Risk Reduction and the Southern Africa Development Community's Disaster Risk Reduction Strategy (SADC, 2018). The strategies emphasize disaster risk management and increased resilience to climate change through increased political commitment to disaster risk reduction, improved identification and assessment, enhanced knowledge management, increased public awareness, and improved governance of risk (UNDP, 2014b). Public awareness and education on climate change are the foundation of the success of disaster risk management initiatives and have improved efforts on risk analysis and early detection mechanisms (World Bank Group, 2021).

Botswana Adaptation Strategy in the Context of Climate Justice

The global response to climate change is embodied in the United Nations Framework Convention on Climate Change (UNFCCC), article 2's objective as:

> to achieve stabilization of greenhouse gas concentrations in the atmosphere at levels that would prevent dangerous anthropogenic interferences with the climate system allowing ecosystems to adapt naturally, ensuring that food production is not compromised and at the same time, allowing development to continue sustainably.

Botswana ratified both the Convention and the Protocol in 1994 and 2004, respectively (UNDP, 2014b). The country has succeeded in coming up with a framework that is in line with global expectations and the climate justice framework. According to Mary Robinson Foundation (2019):

> Climate justice links human rights and development to achieve a human-centred approach, safeguarding the rights of the most vulnerable people and sharing the burdens and benefits of climate change and its impacts equitably and fairly. Climate justice is informed by science, responds to science, and acknowledges the need for equitable stewardship of the world's resources.

The principles include the respect and protection of human rights, support for the right to development which emphasizes resource fairness between countries in the North and South, sharing of benefits and burdens equitably, decisions on climate change are participatory, transparent and accountable, highlight gender equality and equity, harness transformative power of education for climate stewardship, and the use of effective partnerships to secure climate justice (Mary Robinson Foundation, 2019). Botswana is committed to allocating resources for climate change research and further collaborates with institutions of learning at national, regional, and international levels to promote adaptation and mitigation-related research (UNDP, 2017).

Botswana has realized the importance of including marginalized groups in its adaptation strategy as well as its collaboration with regional and international bodies in its endeavour to mitigate the effects of climate change. The success of such strategies remains to be seen as the country still relies heavily on the burning of fossil fuels to provide energy. The overreliance on coal and coal products for the production of electricity at Morupule by the Botswana Power Corporation cannot be abandoned overnight. These challenges will be elaborated on further under the next subtopic. Focusing on the Concern Worldwide (2022) framework on climate justice and injustice as presented above, Botswana's adaptation makes ample room for the consideration of issues such as geographical inequalities, marginalized people, youth, women, and the poor (agriculture).

The National Adaptation Policy of the Government of Botswana (Ministry of Environment, Natural Resources Conservation and Tourism, 2020) consider the involvement of the rural and urban areas in the planning for climate change adaptation strategies which is an important step in bridging the geographical inequality disparity. Again, the same document promotes the leveraging of the private sector to equip farmers with technologies such as climate-smart agriculture, a strategy to capture the marginalized groups who are most vulnerable to climate change's impact. Considering that women in Botswana work more in the natural resource-reliant sectors, the national adaptation policy makes provision for gender-responsive and human rights approaches. The plight of the youth is also captured in the national adaptation policy of Botswana through youth-centred approaches as the youth will bear most of the burden of climate change over a long period. On poverty, the national adaptation policy of Botswana considers a pro-poor approach which also covers vulnerable groups such as people with disabilities, elderly people, and severely ill people within the society.

Challenges of Climate Injustice in Botswana

Climate change in Botswana is deepening inequalities between men and women in agriculture (Kunjinga, 2014). Women are mostly involved in crop production while their male counterparts focus on cattle production. Due to the erratic rains and frequent droughts, crops usually fail. During times of drought, women have the unimaginable task of providing food for their families. Since 1980, the region has faced climate change effects such as the reduction of rainfall, frequent droughts, and flooding. At the community level, climate change adaptation has failed due lack of knowledge about climate change and adaptation strategies, weak local institutions, and a lack of material and financial resources needed for enhancing and measuring the effectiveness of adaptation strategies.

The technology needed to combat climate change is quite expensive for middle-income countries such as Botswana which are found predominantly in the global South. The long heat periods that part of Botswana is exposed to are a vulnerability to those communities but could be used as an advantage if the government were to invest in solar energy. The abundance of sunlight could be used to reduce the reliance on fossil fuels and gas unfortunately funding is currently scarce to establish such a monumental power station in Kgalagadi district as this would be an ideal part of the country.

The countries of the global North have a responsibility to assist those to the South in acquiring such technology since they have benefitted tremendously through industrialization the driving force behind climate change and global warming. This is however not the case as they too still depend on fossil fuels and are reluctant to assist developing countries in cutting their carbon emissions by investing in green technology. The idea of an electric car for every home is also hindered by the technology that could be required to charge the batteries used by these vehicles every home would have to be fitted with a charging system. On average a battery takes up to 3 hours to charge therefore it would be challenging if people would have to queue to access a charge (Guo and Kontou, 2021).

Conclusion

Climate change strategies can only be affected if they are embodied in the national policy-making agenda. Departments within certain ministries such as the ministries of agriculture, lands, energy, and tourism should be mandated with spearheading climate change strategies. Acts of parliament should be passed that make climate change adaptation strategies law and that they are adhered to fully. The participation of marginalized groups women and the youth, civil society organizations and the government are key when formulating policies and if any attempt at mitigating climate change can be successful. The international community should take the lead in ensuring that resource injustices are alleviated between the North and the South and they must ensure equity as well as provide funding to less developed countries to acquire and train experts in green technology. It is worth noting that the national adaptation policy of Botswana is very mindful of the issues of climate justice as outlined in this study.

References

Adaptation at Scale in Semi-Arid Regions (ASSAR) (2018). Adapting to climate change in semi-arid Botswana: ASSAR's key findings. Available at http://www.assar.uct.ac.za/sites/default/files/image_tool/images/138/Botswana/ASSAR%20-%20Botswana%20final%20stakeholder%20report%20-%20November%202018%20-%20web%20version.pdf. (Accessed 5th July, 2023).

Adedeji, O., Okocha, R., and Olatoye, O. (2014). Global climate change. *Journal and Environment Protection* 2: 114–122. (Online) Available at: http://www.scirp.org/journal/gep http://dx.doi.org/10.4236/gep.2014.22016. (Accessed 24th Jan. 2023).

Baxi, U. (2016). Towards a climate change justice theory?, Journal of Human Rights and the Environment, 7(1), 7–31. https://doi.org/10.4337/jhre.2016.01.01. (Accessed 6th July 2023).

Behrens, A., Georgiev, A., and Carraro, M. (2010). Future impacts of climate change across Europe. *Centre for European Policy*. (Online) Available at: http://aei.pitt.edu/14586/1/WD_324_Behrens,_Georgiev_&_Carraro_final_updated_(1).pdf. (Accessed 20th Dec. 2023).

Boko, M., Niang, I., Nyong, A., and Vogel, C. (2007). Climate change adaptation among smallholder farmers: Evidence from Ghana. (Online) https://www.scirp.org/%28S%28vtj3fa45qm1ean45vvffcz55%29%29/reference/referencespapers.aspx?referenceid=2852593. (Accessed 6th Jan. 2023).

Chaudhary, A., and Mooers, A. (2018). Terrestrial vertebrate biodiversity loss under future global land use change scenarios. (Online) https://doi.org/10.3390/su10082764. (Accessed 10th Feb. 2023).

Concern Worldwide (2022). Climate Justice, Explained. Available at https://www.concern. net/news/climate-justice-explained. (Accessed 5th July 2023).

Chaudhary, A., and Mooers, A. (2018). Terrestrial vertebrate biodiversity loss under future global land use change scenarios. (Online) https://doi.org/10.3390/su10082764. (Accessed 10th Feb. 2023).

Dolšak, N., and Prakash, A. (2022). Three faces of climate justice. *Annual Review of Political Science* 25:283–301. https://doi.org/10.1146/annurev-polisci-051120-125514. (Accessed 28th Dec. 2023).

Frimpong, O.B. (2020). Climate Change and Violent Extremism in the Lake Chad Basin: Key Issues and Way Forward. https://www.wilsoncenter.org/publication/climate-change-and-violent-extremism-lake-chad-basin-key-issues-and-way-forward. (Accessed 5th July 2023).

Guo, S., and Kontou, E. (2021). Disparities and equity issues in electric vehicles rebate allocation. *Energy Policy* 154:112291. https://doi.org/10.1016/j.enpol.2021.112291. (Accessed 24th Jan. 2023).

Government of Botswana. (2001). Botswana: Climate risk country profile. (Online) https://reliefweb.int/report/botswana/botswana-climate-risk-country-profile. (Accessed 12th Jan. 2023).

Homberg, M.V., and McQuistan, C. (2019). Technology for climate justice: A reporting framework for loss and damage as part of key global agreements. In: Mechler, R., Bouwer, L., Schinko, T., Surminski, S., and Linnerooth-Bayer, J. (eds) *Loss and Damage from Climate Change. Climate Risk Management, Policy and Governance*. Cham: Springer. https://doi.org/10.1007/978-3-319-72026-5_22. (Accessed 20th Jan. 2023).

IPCC. (2007). Working group "climate change 2007: Working group II: Impacts, adaptation and vulnerability". Chapter 9: *Africa*, pp. 433–467. (Online) Available at: https://archive. ipcc.ch/publications_and_data/ar4/wg2/en/ch9.html. (Accessed 23rd Jan. 2023).

IPCC Report. (2018). Third assessment report of the intergovernmental panel on climate change. (Online) https://unfccc.int/third-assessment-report-of-the-intergovernmental-panel-on-climate-change?gclid=CjwKCAiAl9efBhAkEiwA4TorilmsWGF6yXo2ZmCM3TncTyjcKR7Hlqg153_MDsjO8QwTrCO5u0pJpBoC5FAQAvD_BwE. (Accessed 10th Feb. 2023).

Joshua, D. (2021). The Sahel in the midst of climate change. (Online) https://reliefweb.int/report/chad/sahel-midst-climate-change. (Accessed 9th Feb. 2023).

Kgathi, D.L. (2018). Ecosystem services and human well-being at lake Ngami, Botswana: Implications for sustainability. (Online) https://novapublishers.com/shop/ecosystem-services-and-human-well-being-at-lake-ngami-botswana-implications-for-sustainability/. (Accessed 20th Feb. 2023).

Ki-moon, B. Secretary-general of the United Nations (UN). (2009). UN Headquarters, New York. (Online) Available at: https://bankimooncentre.org/tag/climate-change. (Accessed 22nd March 2021). (Accessed 14th Jan. 2023).

Kujinga, K. (2014). Climate change related gender inequality and challenges to community-based adaptation in Ngamiland, Botswana. (Online) https://www.weadapt.org/sites/weadapt.org/files/legacy-new/placemarks/images/original/555c4d227e6e317005173097-6e04b985f8-o.jpg. (Accessed 10th Feb. 2023).

JRF. (2014). Climate change and social justice: Evidence review. (Online) https://www.jrf. org.uk/report/climate-change-and-social-justice-evidence-review. (Accessed 26th Jan. 2023).

Levy, B.S., and Patz, J.A. (2015). Climate change, human rights, and social justice. *Annals of Global Health* 81(3): 310–322. https://doi.org/10.1016/j.aogh.2015.08.008

Mary Robinson Foundation-MRF. (2019). Principle of climate change. (Online) https://www.mrfcj.org/principles-of-climate-justice/ (accessed 27th Jan. 2023).

Ministry of Environment, Natural Resources Conservation and Tourism. (2020). National Adaptation Plan Framework for Botswana. Available at https://napglobalnetwork.org/wp-content/uploads/2020/06/napgn-en-2020-nap-framework-for-botswana.pdf. (Accessed 5th July 2023).

Ministry of Environment, Wildlife and Tourism (MEWT). (2012). Second national communication to the United Nations Framework Convention on Climate Change (UNFCCC). Retrieved from UNFCCC website: http://unfccc.int/resource/docs/natc/bwanc2.pdf. (Accessed 25th Jan. 2023).

Nkemelang, T., New, M., and Zaroug, M. (2018). Temperature and precipitation extremes under current, 1.5 °C and 2.0 °C global warming above pre-industrial levels over Botswana, and implications for climate change vulnerability. (Online) https://iopscience.iop. org/article/10.1088/1748-9326/aac2f8. (Accessed 10th Feb. 2021).

Notre Dame Global Adaptation Index (ND-GAIN). (2020). Botswana profile. Retrieved from http://index.gain.org/country/botswana. (Accessed 25th Jan. 2023).

Republic of Botswana. (2019). Botswana's First Biennial Update Report (Bur) To the United Nations Framework Convention on Climate Change. A document Prepared by the Ministry of Environment, Natural Resources Conservation and Tourism of Botswana.

SADC. (2018). Environment & climate change. (Online) https://www.sadc.int/pillars/environment-climate-change. (Accessed 10th Feb. 2023).

Schlosberg, D. (2012). Climate Justice and Capabilities: A Framework for Adaptation Policy. *Ethics & International Affairs*, 26(4), 445–461. doi:10.1017/S0892679412000615

Schlosberg, D., and Collins, L. (2014). From environmental to climate justice: Climate change and the discourse of environmental justice. *WIREs Climate Change* 5(3):359–374.

Skah, M., and Lyammouri, R. (2020). The climate change-security nexus: A case study of Lake Chad. *Policy Centre for the New South*, pp. 1–36. (Online) Available at: https://media.africaportal.org/documents/the_climate_change_security_nexus.pdf. (Accessed 19th Jan. 2023).

Southern African Development Community (SADC). (2018). Botswana national climate change strategy 2018 – SADC. (Online) https://drmims.sadc.int/sites/default/files/document/2020-03/2018_Botswana%20Climate%20Change%20Strategy.pdf. (Accessed 20th Feb. 2023).

Sovacool, B.K., Ali, S.H., Bazilian, M., Radley, B., Nemery, B., Okatz, J., and Mulvaney, D. (2020). Sustainable minerals and metals for a low-carbon future. *Science* 367(6473):30–33. https://doi.org/10.1126/science.aaz6003. (Accessed 25th Jan. 2023).

Stern, N. (2006). *The Economics of Climate Change—The Stern Review.* Cambridge: Cambridge University Press.

Tadesse, D. (2010). Impact of climate change in Africa. *Institute for Security Studies (ISS Paper)* (220), pp. 2–20. (Online) Available at: https://www.files.ethz.ch/isn/136704/PAPER220.pdf. (Accessed 24th Jan. 2023).

University of Gothenburg. (2008). Botswana Environmental and Climate Change Analysis 29 May, 2008. Available at https://sidaenvironmenthelpdesk.se/digitalAssets/1683/1683296_environmental-and-climate-change-policy-brief-botswana-2008.pdf. (Accessed 5th July 2023).

UNICEF. (2018). Climate change poses significant risks to children's health and well-being. (Online) https://data.unicef.org/topic/climate-change/overview/. (Accessed 12th Feb. 2023).

UNDP. (2017). Key highlights of UNDP work in Botswana 2017–2021. (Online) https://www.undp.org/botswana/news/key-highlights-undp-work-botswana-2017-2021. (Accessed 5th Jan. 2023).

United Nations Framework on Convention Climate Change (UNFCCC). (2015). (Online) https://unfccc.int/process-and-meetings/what-is-the-united-nations-framework-convention-on-climate-change. (Accessed 12th Feb. 2023).

United Nations (UN). (2015). The Millennium Development Goals Report 2016. Available at https://www.un.org/millenniumgoals/2015_MDG_Report/pdf/MDG%202015%20 rev%20(July%201).pdf. (Accessed 5th July 2023).

United Nations Development Programme (UNDP). (2014a). Human development report 2014: Sustaining human progress: Reducing vulnerabilities and building resilience. Retrieved from http://hdr.undp.org/en/2014-report. (Accessed 25th Jan. 2023).

United Nations Development Programme (UNDP). (2014b). Botswana climate change draft 2. (Online) https://info.undp.org/docs/pdc/Documents/BWA/DRAFT%20CLIMATE%20 CHANGE%20RESPONSE%20POLICY%20%20version%202%20(2).doc (Accessed 20th Feb. 2023).

WHO. (2015). Climate change and health. (Online) https://www.who.int/news-room/fact-sheets/detail/climate-change-and-health. (Accessed 13th Feb. 2023).

World Bank Group. (2021). World bank group climate change action plan 2021–2025: Supporting green, resilient, and inclusive development [EN/AR/ZH]. (Online) https://reliefweb.int/report/world/world-bank-group-climate-change-action-plan-2021-2025-supporting-green-resilient-and?gclid=CjwKCAiAl9efBhAkEiwA4TorirOpvWxhxzP0rid4YhlXogsVcvy Q-XtNGkAUd37rT1MLFEjCMXPdtBoCy60QAvD_BwE. (Accessed 12th Feb. 2021).

Part III
Climate Mobility

12 The Climate Change–Migration Nexus

Climate Mobility and Climate Justice in Matabeleland, Zimbabwe

France Maphosa

Introduction

Climate change and migration, particularly irregular migration, are growing and severe global challenges, leading to the suffering, displacement, and death of millions of people worldwide. Climate change leads to increased natural disasters such as floods, mudslides, and droughts which lead to the displacement and death of people. The 2015 Paris Agreement (United Nations, 2015) stresses that climate change represents an urgent and potentially irreversible threat to human societies and the planet. At the same time, migration is increasing. In 2017, Peter Maurer, the president of the International Committee of the Red Cross (ICRC) stated that migration and internal displacement are currently among the most pressing topics on the international agenda. He pointed out that at that time there were 250 million people globally who were living outside their home countries due a diversity of factors including lack of employment opportunities in their countries of origin, scarcity of health and educational services, and the desire to join family members already in other countries. According to the International Organisation for Migration (IOM) (2020), the number of international migrants globally is 27.2 million. The United Nations High Commissioner for Refugees (UNHCR) (2020) reported that in 2019, the number of forcibly displaced people globally was 79.5 million. Out these, 26 million were refugees and 47 million were internally displaced people.

According to the United Nations (2019) migration and climate have always been connected. Piguet (2011) states that early migration theorists such as Ravenstein acknowledged climate as a driver of migration. He points out that despite these early insights, climatic conditions as a driver of migration gradually disappeared from migration literature. At the same time, the United Nations (2019) cautioned that the impacts of the man-made climate crisis are likely to extensively change the patterns of human settlement. Bezu et al. (2020) state that it is predicted that in the coming decade, climate change will contribute to the migration of tens of millions of people. While some migration will be caused by a sudden onset of events such as floods, most of it will be due to the long-term impact of climate change on livelihoods. Countries that depend on environmental resources and vulnerable ecosystems for livelihood are the most affected.

DOI: 10.4324/9781003397120-15

While there is general agreement that climate factors contribute to migration, the consequences of climate change on migration are not straightforward. Establishing a linear causal link between climate factors and migration has not been possible. To resolve this difficulty, scholars have adopted the concept of "climate mobility" in linking climate change with migration. The concept of climate mobility acknowledges that migration under climate change takes different, contextually dependent forms shaped by existing relations including inequality (Boas et al., 2022). Climate mobility therefore presents climate related migration as a selective process which is affected by differential and unequal distribution of resources which affects an individual or group's to adaptation climate change, migration being one of the forms of climate adaptation (Wulf and Newton, 1996). The concept of climate mobility therefore links migration with issues of social and climate justice.

While Africa has been affected by the global challenges of climate change and migration, there are not many studies that link these two challenges. This chapter employs the concept of climate mobility to link human mobility in Matabeleland to climate change. It argues that while climate factors contribute to human mobility, they are compounded by pre-existing socio-economic and political conditions of social injustice. Climate change exacerbates the impacts of inequalities on the communities existing in that region.

Climate Change and Migration in Africa

The Intergovernmental Panel on Climate Change (IPCC) reports that despite its small contribution to climate change, Africa is the continent that will be most affected by it (Portner et al., 2022). Tadesse (2010) also points out that although African nations are the least polluters, analysts envisage that they will suffer most from it. In fact, the African continent is already experiencing the impacts of climate change. Climate change has devastating effects on agricultural production, food security, shelter, people's livelihoods, and economic development, particularly in many regions of the African continent (United Nations, 2017). The United Nations' Food and Agriculture Organisation (FAO) predicted that in some of Africa's poorest regions, the impact of climate change would lead to as much as 50% fall in crop yields by 2020. The United Nations (2020) affirms that climate change is an increasing threat to Africa. The World Meteorological Organisation(WMO) has urged that there is need for urgent efforts to be pursued to address the agricultural, health, and economic implications of climate change in Africa. Tadesse (2010) asserts that climate change is also a security threat in Africa, as environmental degradation and demographic pressures resulting in the displacement of millions of people create social upheaval. Africa's Agenda 2063 also recognises the challenges posed by climate change to development.

Migration within and out of Africa has a long history (Shimeles, 2010; IOM, 2020). The IOM (2020: 4) states that "Africans have always and will continue to migrate in search of opportunities and sometimes safety". According to Shimeles (2010), the estimated number of people of African descent living outside the continent is approximately 140 million. As observed by Akokpari (2000), migration in Africa is largely informal and this makes it extremely difficult to document. As a result, accurate statistics on the phenomenon of migration in Africa, both inside and outside the continent, are scarce. Despite the scarcity of data, there is an indication that migration in the continent is on the increase. This migration is largely forced migration because it is caused by a combination of factors including conflicts, political oppression, economic crises, and environmental factors. Akokpari (2000: 72) argues that that these factors are being reinforced by globalisation, creating a "… huge and seemingly unmanageable migration problematic on the continent". Globalisation has not only eased the mobility of people across national borders but has also enabled them to maintain connections with their communities of origin, resulting in transnationalism.

The IOM (2020) states that most literature on African migration is by writers from the West. It argues that there is limited migration scholarship on African migration by African scholars (IOM, 2020). IOM argues that the few African scholars on African migration are largely influenced by the hegemonic Western thinking which ultimately influences the African policy makers. The organisation therefore advocates for the change of the migration narrative in Africa. While this might be valid for historical literature, recently, there has been a surge in the number of publications on migration in Africa by African scholars (Sachikonye, 1988; Zinyama, 1990; Mbiba, 2005; Maphosa, 2007; Zanamwe and Devillard, 2009; Maphosa, 2011; Thebe, 2011; Maviza, 2020; Maphosa and Ntau, 2020).

Migration and Climate Change in Southern Africa

Intra-regional migration within the Southern African region dates back to the mid-19th century (Crush et al., 2006), South Africa being the most popular destination. Migration from other countries to South Africa increased significantly after the abolition of apartheid in 1994 which resulted in the opening up of the country to outsiders. South Africa continues to attract immigrants from other countries in the region because of its middle-income status, its stable democratic institutions and comparatively industrialised economy (Moyo, 2021). A combination of social, political, economic, and environmental factors have been identified as contributing to migration within the region (World Bank, 2018). These include seeking for employment, re-uniting with families, and fleeing internal conflict (UN, 2022). As observed by Oucho (2007) migration in Southern Africa is therefore a combination of forced and voluntary migration. It is also a combination of regular and irregular migration. There are no reliable statistics on irregular migration which is mainly facilitated by human smuggling from which criminals generate huge profits while exposing those dependent on smugglers to exploitation and sexual abuse (UN, 2022).

Climate change has, however, not featured prominently among the factors motivating people to migrate within the region. This is despite the region being vulnerable to the impact of climate change leading to, among other things, food insecurity (Young et al., 2010). Mbiyozo and le Roux (2021) observe that Southern Africa is the region that is worst affected globally by climate change, despite it contributing little to it. The devastating floods that have left many people dead or displaced and infrastructure damaged in the region over the last few decades have been attributed to climate change (Wamukonya and Rukato, 2001). The World Food Programme (WFP) (2021) observes that climate change is a long-term threat to food security and nutrition in the region. It reports that in the past decade the region has experienced a deficit in cereal production from 0.1 to 8.9 million metric tonnes. This is because climate hazards adversely affect the region's predominantly rain-fed agricultural sector (Gosling et al., 2020). Despite the many effects of climate change in the region Mbiyozo and le Roux argue that the region's governments have not prioritised adaptation measures, in a region that has experienced more than 600 climate related disasters since 1980. The need for scholarship linking the migration to climate change is therefore urgent. While mitigation and adaptation and mitigation efforts may be constrained by limited financial, institutional and human resources, evidence-based preventive actions to address climate change–mitigation are crucial (UN, 2017).

Climate Change in Zimbabwe

Predictions are that in Zimbabwe, climate change which is already occurring will cause average temperatures to rise by about 3° before the end of the century (Brazier, 2015). Annual rainfall is predicted to decline by between 5% and 18% especially in the Southern parts of the country. Rainfall will become more variable with an increase in droughts, floods, and storms. These changes will affect the country's food security, health and energy supply, and the economy. The country experienced severe droughts in the 1991–1992, 1994–1995, 2002–2003, 2015–2016, and 2018–2019 agricultural seasons (Frischen et al., 2020). In 2016, the government of Zimbabwe declared a state of disaster in response to a drought that left an estimated quarter of the population without food. These conditions are likely to cause the land to be increasingly marginal for agriculture and this poses a serious threat to the economy and the livelihoods of Zimbabweans especially the poor because of Zimbabwe's heavy dependence on rain-fed agriculture and sensitive resources. The country has already been experiencing droughts which have become more frequent over the last two decades.

As records show that Zimbabwe is already experiencing climate change effects, mainly rainfall variability and extreme events, the government claims to be taking serious steps to address the effects of climate change. The government of Zimbabwe has signed and ratified of the United Nations Framework Convention on Climate Change (UNFCCC). Its 1996 review of environmental legislation incorporated climate change issues. It also intends to mainstream environmental issues in its development plans (UNPD, 2021).

Climate change has drastically affected rainfall patterns across the globe and Zimbabwe is no exception (Muzerengi and Tirivangasi, 2019). This is partly because of its geographical location. Its location in the tropics makes Zimbabwe vulnerable to shifting rainfall patterns and water resources availability (UNDP, 2021). Brown et al. (2012) also hold the view that Zimbabwe's geographical location makes it vulnerable to the effects of climate change. They point out that Zimbabwe lies in a semi-arid region with limited and unreliable rainfall patterns and temperature variations hence extreme weather events, namely tropical cyclones and droughts have also increased (Brown et al., 2012). Climate change leads to more arid environments for agricultural productivity. Due to their heavy reliance on rain-fed agriculture, the livelihoods of the poor in Zimbabwe are the most vulnerable to climate change. The United Nations Development Programme (UNDP)(2021) cautions that given the heavy dependence of the country on rain-fed agriculture, the absence of natural lakes, the frequent occurrence of droughts in the region and growing population, the potential socio-economic impacts of climate change could be devastating.

The combination of continuous occurrence of droughts with economic calamities has contributed to the reduction in grain production among communal farmers, most of whom are in semi-arid areas and has led to serious food shortages particularly in Matabeleland as a result of erratic rainfall which has negatively affected subsistence farming (Muzerengi and Tirivangasi, 2019). In 2016, the government of Zimbabwe declared a state of emergency in response to a serious drought that caused crop failures across the country leaving many communities vulnerable and food insecure. The drought left approximately 2.5 million people requiring food aid. Muzerengi and Tirivangasi (2019) argue that while food insecurity in Zimbabwe is attributable to many factors including political and socio-economic factors, the most gruesome are the effects of climate change.

The effects of climate change are not uniform across the country because of variations in ecological conditions across the country. Zimbabwe is divided into five agro-ecological zones (natural regions). The demarcation was first made in the 1960s. The regions are differentiated by the average annual rainfall, soil quality, and vegetation which decrease as one moves from Region I to Region V. According to Roth and Bruce (1994), Region IV is characterised by:

> … low rainfall (450–600mm) and is subject to periodic seasonal droughts and severe dry spells during the rainy season. Low and uncertain rainfall make cash cropping risky, except for drought resistant crops and soils with better water retention. Farming systems are suited to livestock production with some intensification possible with drought resistant fodder crops.
>
> (p. 8)

In Region V,

> … rainfall is too low and erratic for reliable production of even drought resistant fodder and grain crops. Included in this region are areas below 900 metres … Farming systems based on extensive cattle and game ranching are best suited to these conditions.
>
> (p. 9)

Matabeleland Province predominantly falls within agro-ecological zones IV and V. In this region, crop production is mainly for subsistence purposes. Even in a good year most households produce just enough for their own consumption. The region has experienced a series of severe droughts over the years resulting in drastic reduction in crop production. As a result, most of the households depend on remittances from migrants to South Africa. Hobane (1999) found that 62% of the adult population in Ward Seven of Mangwe District in Matabeleland South were employed outside the country mainly in South Africa. Climate change is aggravating the already precarious food and nutrition situation of communities in this part of the country.

Through various legislation such as the Land Apportionment Act (1930), Native Land Husbandry Act (1951), Land Tenure Act (1969), and Tribal Trust Lands Act (1979), the colonial government disproportionately allocated between blacks and white, giving a larger and more fertile portion of the land to the whites, while pushing the majority indigenous blacks to agriculturally marginal areas. Gonese et al. (2002: 8) observe that by independence in 1980;

- 42% of the land in Zimbabwe (being marginally productive and drought prone land) was reserved for blacks while;
- 51% (being the more fertile, better watered, more productive and better serviced regions of the country) was technically, if not explicitly reserved for whites.

It was expected, therefore, that land reform would top the development agenda of the post-colonial government by, among other things addressing the historical, racially imbalanced in land ownership. However, the various land reform initiatives did not achieve this purpose. For example the Fast Track Land Reform programme which the government of Zimbabwe embarked on in 2000 ostensibly to decongest the communal areas did not achieve its goal. The commission appointed by the then President Mugabe to review the land distribution programme concluded that with regards to Matabeleland North "...the impact on congestions has therefore been negligible" and that decongestion in Matabeleland South had not been effected. The 1960s classification of the agro-ecological regions of Zimbabwe is currently being re-visited (Mugandani et al., 2012). For example while in 1999 Nyamudeza estimated that Region V constituted 30% of the country, in 2012 Mugandani et al. (2012) estimated that it constituted 32.5% of the country. According to Mugandani et al. (2012), while the number of regions has not changed, their sizes have been extended. This is a consequence of the increasing variability in rainfall which is evidence of climate variability and change.

Matabeleland is one of the most arid regions in the country. Its climate is characterised by relatively short, erratic, and variable rainy seasons and long dry winters. The region receives very low average annual rainfall and experiences periodic seasonal droughts and severe dry spells during the rainy season. The semi-arid conditions in Matabeleland including changing rainfall patterns, frequent droughts and land degradation have severely adversely affected both crop and livestock production in the region. According to Tshuma (2020), Matabeleland Province lost approximately 15,000 cattle during the 2019/2020 drought alone.

Migration from Zimbabwe to South Africa

Emigration from Zimbabwe to South Africa is not a new phenomenon. Zimbabweans migrating to South Africa to work in the mining industry started as early as the beginning of the 20th century (Zanamwe and Devillard, 2009). A considerable amount of literature on migration from Zimbabwe to South Africa now exists. Although there are studies on Zimbabwean migration to other countries such as the United Kingdom, the United States of America, and Australia (Gaidzanwa, 1999; Mbiba, 2005; Chingarande and Maphosa, 2007; Bloch, 2008; Pasura, 2010; Madebwe and Madebwe, 2017) most of the literature focuses on migration from Zimbabwe to South Africa (van Onseln, 1976; Sachikonye, 1988; Zinyama, 1990; Crush and Tevera, 2010; Maphosa, 2007, 2011; Thebe, 2011; Nzima et al., 2016; Maviza et al., 2019; Maphosa and Ntau, 2020; Maviza, 2020; Moyo, 2021). This is because cross-border migration between Zimbabwe and South Africa has a long and complex history (McDonald et al., 2000). As a result, Crush and Tawodzera (2016) assert that the migration corridor between the two countries is also increasingly being of interest to researchers. Several major volumes and numerous journal articles have been published, focusing on a wide variety of migration themes including migration drivers, undocumented migration, brain drain, diaspora engagement, return migration, abuse of migrants' human rights, migrant identities, the working conditions, and livelihood strategies of migrants and migration policy responses of the South African government (Tawodzera and Crush, 2016).

The growing volume of migration literature from Zimbabwe to South Africa demonstrates that migration is both an increasing and complex phenomenon. Despite the bourgeoning literature on migration from Zimbabwe to South Africa, there is still a scarcity of literature linking migration to climate change. Interestingly, in a study carried out by Tawodzera and Crush among Zimbabwean migrants in urban areas of South Africa only 4.2% of them mentioned drought as the reason for migrating. There is however a growing awareness that the negative effects of climate change contribute both directly and indirectly to migration locally, nationally and across borders.

Migration from Matabeleland to South Africa

There is growing literature on migration from Matabeleland, especially to South Africa (Maphosa, 2007; Maphosa, 2011; Thebe, 2011; Maphosa and Ntau, 2020; Maviza, 2020; Matose et al., 2022). These studies refer to several factors as drivers of migration from the region. Among the many reasons given for the migration of people from Matabeleland, climate change has not featured as a major driver. Maphosa (2011), for example, found that the factors that affected the decision to leave Matabeleland for South Africa included poverty, unemployment, political and economic marginalisation, unemployment, poor working conditions, political violence, a culture of migration, and a migratory disposition. Poverty is a multifaceted phenomenon with multiple causes. Certainly, poor and changing climatic conditions are one of the causes of poverty in the region. For example, Hobane (1999) found that because of poor and variable rainfall patterns and poor soils, harvests in some parts of Matabeleland do not last until the next harvest. As a result, people are compelled to seek alternative sources of livelihoods, one of them being migration. In this context, migration should be viewed as adaptation to climate change, by which communities seek to reduce the impacts of climate change (Vinke et al., 2022).

Various scholars have decried the political and economic marginalisation of Matabeleland by the post-colonial governments of Zimbabwe (Darbon, 1992; Alexander et al., 2000; Musemwa, 2006). This marginalisation leads to poor infrastructure such as roads, schools, and hospitals. While unemployment in Zimbabwe as a whole has been worsening over the years, in Matabeleland it has been compounded by the political and economic marginalisation that the region has been experiencing since independence in 1980. The political violence that took place immediately after independence popularly known as *Gukurahundi* forced many people from Matabeleland to flee to South Africa. Migration from Matabeleland to South Africa has also been attributed to the existence of a culture of migration. According to Brittell (1993), a culture of migration develops in a community that has experienced migration for a long time to the extent that migration becomes a way of life for the people in that community. The development of a culture of migration is triggered by many factors such a prolonged economic and political marginalisation. It can also be triggered by the benefits from migration. In Matabeleland, the culture of migration to South Africa is also attributed to the long history of migration between the two countries, as result of the contract labour system, transnational kinship ties, linguistic and cultural linkages, geographical proximity, and the development of migrant networks over time. The culture of migration in Matabeleland is also encouraged by the benefits of migration compared to non-migration (Maphosa, 2011).

Zimbabweans who move to South Africa have been referred to differently as "labour migrants", or "economic refugees". The concepts of "climate" and "environment" are missing in the description of migrants to South Africa from Zimbabwe in general and Matabeleland in particular. Crush et al. (2017) explain migration from Zimbabwe to South Africa as an outcome of the protracted economic and political crisis in the country. They argue that the reasons for migration from Zimbabwe to South Africa are clearly linked to the on-going economic crisis in Zimbabwe. In a study of Zimbabwean migrants in South Africa, they found that 80% of them reported that they migrated to South Africa because they wanted to provide for their families back home. Unemployment was found to be a significant drive of migration to South Africa. Human Rights Watch (2005) estimated that since 2005, about 1.5 million Zimbabweans have fled to South Africa. They were fleeing from political persecution in the form of targeted mass evictions and destruction of homes and livelihoods and from economic destitution due to the collapse of the

Zimbabwean economy. However, Munyoka (2020) has highlighted environmental factors as drivers of migration, in particular irregular migration from Zimbabwe to South Africa. He argues that food insecurity has worsened irregular migration from Zimbabwe. For example, the poor agricultural season in 2019 worsened food insecurity in the country.

In South Africa, Zimbabwean migrants are generally viewed as a problem. They have been accused of taking jobs from locals and contributing to the lowering of wages. The description of migrants as a problem or threat to society places them in the category of *homo sacer* a state of exception which means they exist outside the protection of the law (Maphosa and Ntau, 2020). While they are uprooted from their communities of origin they are, at the same time, not integrated into their country of destination (Maphosa, 2011). They exist on the margins of social, political, cultural, economic, and geographical borders (Downey, 2013). According to Downey (2013: 119),

> Denied access to legal, economic and political redress, these lives exist in a limbo-like state that is largely preoccupied with acquiring and sustaining the bare essentials of life. The refugee, the political prisoner, the disappeared, the 'ghost detainee', the victim of torture, the dispossessed, the silenced, all have been excluded, to different degrees, from the fraternity of the social sphere, appeal to the safety net of the nation state, and recourse to international law.

The South African government's response to migrants from Zimbabwe has been predominantly restrictive, exclusionary, and control oriented. This has involved arresting, detention, and deportation (Machinya, 2019). Irregular migrants are vulnerable to human rights violations in their destinations. They are often discriminated against, exploited and generally marginalised. Despite these experiences, the number of irregular migrants from Matabeleland to South Africa keeps increasing. This indicates the existence of strong push factors that require urgent addressing.

The Challenge of Establishing Causality

As stated by the UN (2019) migration and climate have always been connected. This is reflected in the early theories of migration that mentioned climate as one of the drivers of migration (Piguet, 2011). According to Piguet, while the effects of climate on migration are being presented as new or as part of the future trends, they in fact have a long history. Piquet argues that environmental factors ranked highly in the first systematic theories of migration. For example, Ravenstein (1889) referred to unattractive climate as being responsible for producing migration currents. Despite these early historical insights, reference to the environment as an explanatory factor were to progressively disappear from the migration literature over the course of the 20th century. Piguet (2011) observes that since the end of the eighties there have been numerous theoretical publications on international migration but without any mention of the environment.

Despite the disappearance of climate as an explanatory factor in migration for a while in migration literature, there is now a growing consensus among researchers on the influence of climate on human mobility (ActionAid, 2020; Naser, 2021; Detges et al., 2022; Virgil et al., 2022; Hoffman, 2022). The United Nations Convention to Combat Desertification (UN, 2014) report on desertification states that climate change and desertification are increasingly displacing people and forcing then to migrate. Yet the UN (2017) believes that the strong and often mutually reinforcing interlinkages between climate change, migration, and conflict have not yet received concerted and adequate attention. In 1990, the Intergovernmental Panel on Climate Change (IPCC) noted that the greatest single impact of climate change could be on human migration, with millions of people being displaced by shoreline erosion, coastal flooding, and agricultural disruption (Brown, 2008).

Various analysts have attempted to predict the number of future "climate migrants" or "climate refugees". The well-known analyst Myers (2005) predicted that climate change will produce 200 million climate refugees by 2050. This will be due to global warming which will lead to disruptions of monsoon systems and other rainfall regimes, droughts of unprecedented severity and duration and sea level rise and coastal flooding. Despite being the most cited figure, Myers' estimation of 200 million climate change refugees by 2050 has been challenged by many researchers (Brown, 2008; Bezu et al., 2020). Brown (2008) cautions that repetition does not make the figure any more accurate. Bezu et al. (2020) are critical of this figure for what they refer to as its implication that all those vulnerable to climate change, including the youth, will migrate, which is not necessarily the case. They argue that in addition to climate change, a multitude of other factors influence the decision to migrate and these other factors include personal, political, economic, and environmental conditions. Brown (2008) also argues that disaggregating the role of climate change from other factors requires an ambitious analytical step into the dark.

Scientific evidence on climate change is widely being acceptable. However, the consequences of climate change for human migration are not straightforward. This is because there many other social, economic, and environmental factors involved in the decision to migrate that make establishing a linear causative relationship between anthropogenic climate change and migration difficult (Brown, 2008). The relationship between climate change and migration is a complex one and several writers have tried to tackle it (Tacoli, 2009; Warner, 2010, Black et al., 2011; Bezu et al., 2020). The effect of climate change on migration has also become of increasing interest for both policymakers and researchers yet, according to Piguet (2011) knowledge on the subject remains limited and fragmented. The task of empirically linking climate change and migration is not straightforward (Bezu et al., 2020). There are various other non-climate factors such as government policy, population growth and community level resilience to natural disasters that also contribute to the degree of vulnerability people experience and that affect the decision to migrate (Brown, 2008). Migration is not just a result of a decision, it is also a result of the ability to migrate. The ability to migrate is influenced by availability of resources which can be financial or social. This means, as Brown observes, that the people most affected by climate change are not necessarily the ones most likely to migrate (Brown, 2008). These include the poorest of the poor, the young and women.

For Nicholson (2011), to say that the environment affects migration is a truism. What is important is for research to establish the precise nature and dynamics of this relationship. This should then lead to an understanding that can help guide policy making. Since climate change is both a direct and indirect driver of migration, the relationship between environmental change and migration is therefore complicated by the contribution of other factors such as population growth, poverty, governance, human security, and conflict. Migration is just one of the many adaptive solutions for those affected by climate change (IOM, 2020). The existence of a multiplicity of other factors that affect the decision and ability to migrate has led some people like Nicholson to ask the question of whether or not the climate change–migration nexus is an analytically meaningful subject for research (Nicholson, 2011). Nicholson contends that the environmental-migration nexus "has no intrinsic capacity to provide us an operable understanding that can guide policy work" (Nicholson, 2011). To understand the role of climate change in migration requires the analysis of how and why people are vulnerable to climate change. It also requires the analysis of the different strategies they develop to adapt to climate change.

These sentiments are shared by the UN (2017) which argues that climate change is not the only factor pushing people to leave their homes, but migration is likely to take place in areas where the local, national, and regional capacities are weak and communities are already under stress due to other factors such as poverty, high unemployment, and food insecurity. In such circumstances climate change may be a compounding driver of forced migration. Sherbinin (2020) concurs with this thinking by stating that human beings possess both the ability to adapt and freewill. One therefore must be careful to avoid environmental determinism that draws a direct line from projected climate changes to future migration. Sherbinin argues that instead, climate circumstances are one of the several factors that drive the decision by an individual or community to migrate.

Waldinger and Fankhauser (2015) point out that evidence shows that people in developing countries are likely to respond to climate change by migrating internally and there is less evidence on the relationship between climate change and international migration. They emphasise that the effect of climate change on migration depends on socio-economic, political, and institutional conditions. These conditions affect both vulnerability to climate change and how important climate change is in determining migration decisions. Waldinger and Fankhauser (2015) further argue that migration might be an effective response to the climate risks of the future only under certain preconditions including access to information on the economic costs of migration and on the advantages and disadvantages of potential destination locations. This information can help potential migrants to make decisions that will improve their livelihoods. Brown (2008) argues that because individual migrants' reasons to leave their homes vary so widely, establishing causality between economic "pull" factors and environmental "push" factors is often subjective.

The disaggregation of the role of climate change from other factors therefore requires an innovative approach. Bezu et al. (2020) advocate for a framework that should focus on how climate change influences the already acknowledged drivers of migration. For Sherbinin (2020), climate should be considered as an envelope in which all economic activities take place. Scientific evidence of climate change is overwhelming. However, the numbers of people who migrate in response to climate change are not yet certain. Brown (2008) believes that the estimates of the numbers of people who will in future migrate in response to climate change are repeated either for shock value or for lack of a better figure. He argues that while the relationship between climate change and migration is unpredictable, not much time, energy and resources, however, have been spent on empirical analysis of the impacts of climate change on human migration.

The Problem of the Terminology

Another issue relating to the relationship between climate change and migration is the terminology used in describing those who migrate because of climate change. According to IOM (2014), environmental (or climate) migration is an emerging field of study and therefore specific terminology is still being developed. However, Brown (2008) contends that labels are important. This is because the way migrants are labelled influences their treatment especially by the international community. The way they are labelled has implications for the obligations of the international community under international law. A migrant is often perceived as an individual who exercises choice to move, usually in search of the proverbial "greener pastures". A refugee, on the other hand, is one who is forced to move because of unbearable circumstances in the country or place of origin. Forced migration entails human movement where an element of coercion exists, including threats to life, and livelihoods, whether arising from natural or man-made causes (UN, 2017). The status of a refugee therefore elicits more sympathy than that of a migrant. In fact, migrants may be treated with hostility in their countries of destination as they may be perceived to be taking away jobs and other benefits from the locals. The status of refugee implies duress and carries fewer negative connotations than that of a migrant which implies voluntary movement towards a more attractive lifestyle (Brown, 2008). Despite the threat to life and livelihoods that climate change poses, there is considerable resistance to the idea of expanding the definition of refugees to include climate refugees. As a result, there is no home for climate migrants in the international community, both literally and figuratively (Brown, 2008). According to Podesta (2019), there is currently no multilateral strategy or legal framework which considers climate change as a driver of migration. Due to the worsening climate conditions, there is definitely going to be an increase in "climate migrants". These people, however, move with little legal protection because the current international law does not recognise them as such. Ionesco (2019) gives ten reasons why those who move because of climate change should not be referred to as refugees. Among the reasons she gives are that climate migration is mainly internal, migration as a result of climate change is not necessarily forced and that it is difficult to isolate climate reasons for migration.

Climate Mobility

Climate mobility is a concept that acknowledges that the impacts of climate change contribute to migration (Szaboova and Colon, 2020) without seeking to establish a causal relationship between climate change and migration. Boas et al. (2022) point out that migration scholarship has moved beyond linear and exceptional terms but that migration under conditions of climate change is contextual and is affected by power relations and inequality. In other words, climate change does not have a direct impact on migration but operates through socio-economic factors such as wage differentials, family re-unification, and the quest for improved living standards (Stojanov et al., 2021). Migration is therefore considered as a form of climate change adaptation (Vinke et al., 2022). As a result, migration in the context of climate change does not constitute one singe act but a wide range of mobilities such as short-term displacements, long distance migration, and circular migration (Boas et al., 2022). It also involves immobilities. This is because some people affected by climate change may face a variety of constraints to leaving, including lack of resources to do so (Wulff and Newton, 1996; Ornstein, 2017; Cook and Butz, 2019). To these "trapped populations", migration is not an option, even in the face of climate related threats (The White House, 2021). In Zimbabwe, the impacts of climate change are not evenly distributed (Chanza and Gundu-Jakarasi, 2020; Bhatasara and Nyamwanza, 2022). Bhatasara and Nyamwanza emphasise that:

> Those that are located in fragile ecosystems, those in rural, agrarian and climate dependent economies, the poor, women and children are affected first and worst.
>
> (Bhatasara and Nyamwanza, 2022)

Climate Mobility, Social Justice, and Environmental Justice

Discourse on climate mobility gives rise to issues of social justice and human rights (Farbotko et al., 2022). In fact, as Farbotko et al. (2022) argue, climate mobility can represent the existence of a significant injustice and human rights challenge. Social justice refers to a fair and equitable distribution of resources, opportunities and privileges in society (Mollenkamp, 2022). Rawls (1971) enunciated the principles of social justice as access, equality, diversity, participation and human rights. The principle of access states that the resources available to society should be accessible to all. Equality means that people should be afforded the same opportunities to succeed. Diversity relates to leadership; that leaders in government or business should be representative of the community they serve. Participation requires that everyone within the community should be given a voice in making decisions. The fundamental principle of social justice is human rights. It recognises that human beings have rights that should be upheld at all times. All these principles are intended to achieve fairness in the allocation of social goods. How social goods are distributed in society should be based on equal rights, equal opportunity, and equal treatment (United Nations, 2006). Social justice therefore requires both distribution and recognition (Frazer, 1998). Social justice influences mobility justice, where issues of power, social exclusion, and unequal access to social goods affect the ability and the nature of movement. As a consequence of the socio-economic disparities caused by social injustice, the impacts of climate change are not borne equally or fairly between the rich and the poor, men and women, the old and the young (Concern Worldwide, 2022). They are also not borne equally between and within countries. The impacts of climate change are felt more by those who contribute the least to it. As observed by Stephens (2007) it is now widely acknowledged that much of the burden of environmental ill-health worldwide falls on the poorer people. Environmental justice is therefore integral to social justice and guarantees that all people should have equal access to a healthy, safe, and sustainable environment and equal protection from environmental harm (Agyeman et al., 2016). Environmental justice therefore promotes a fair distribution of resources in order to address the impacts of climate change. Addressing the causes and impacts of human mobility in the context of climate change necessitates addressing the issue of environmental justice by ensuring through government action in collaboration with other stakeholders including the vulnerable communities. At the core of this is the dealing with the inequitable distribution of environmental risks. As a form of adaptation (Stojanov et al., 2021), climate mobility is shaped by, among other things, the ability to move away from the climate risk situation (Boas et al., 2022). That is why migration under climate change takes many different forms. Guaranteeing environmental justice requires consultation with individuals and communities vulnerable to climate change to ensure equity and inclusion (The White House, 2021).

Social Justice and Climate Mobility in Matabeleland

Zimbabwe, like the rest of the world, is experiencing the impacts of climate change. However, due to its heavy dependence on rain-fed agriculture and sensitive resources, it is particularly vulnerable to climate change (Chikodzi et al., 2013; Chanza and Gundu-Jakarasi, 2020; Frischen et al., 2020; Mushore et al., 2021; Bhatasara and Nyamwanza, 2022). According to Chanza and Gundu-Jakarasi (2020), evidence of climate change in Zimbabwe includes declining water resources, fall in agricultural activity, decline in biodiversity, and the spread of vector-borne diseases, among others. Climate change impacts are not uniform across the country's different provinces (Chanza and Gundu-Jakarasi, 2020; Hunter et al., 2020; Bhatasara and Nyamwanza, 2022). For example, Bhatasara and Nyamwanza (2022: 187) observe that:

> … people who are located in fragile ecosystems, those in rural, agrarian and climate dependent economies, the poor, women and children are being affected first and worst.

A combination of poor soils, poor rainfall patterns and decades of political and economic marginalisation makes Matabeleland to be the region most affected by climate change in Zimbabwe (Madurga et al., 2021). Due to unfavourable climatic conditions, crop harvests in Matabeleland are so low that they hardly last until the next harvest (Hobane, 1999). In a study in Matabeleland North,

Chingarande et al. (2020) observed the production of sorghum, millet, and maize is so marginal that nothing reaches the markets. In fact, the study found that the total household harvest of maize, sorghum, and millet lasts for only four months. Ndlovu et al. (2020), for example, state that Matabeleland South which used to be a thriving livestock and small grain producing province is now relying on humanitarian assistance from the government and other humanitarian organisations. However, Madurga et al. (2021) point out to the partisan nature of the allocation of food aid. They argue that food insecurity and malnutrition in parts of Zimbabwe are a result of patronage, rent seeking behaviour and competition for resources. This is because as Ndhlovu (2021) asserts, Zimbabwe's post-colonial government has used development to marginalise and exclude minority ethnicities from its development agenda. The impact of climate change in Matabeleland is exacerbated by the existing conditions of social injustice, characterised by unfair an inequitable division of resources, opportunities and privileges in the country (Mollenkamp, 2022).

Studies show that generally, climate mobility is internal (Stojanov et al., 2021; The White House, 2021). However, in the case of Matabeleland, most of the movement of people from the region is cross-border movement especially to South Africa. This is because of several factors including the geographical proximity of Matabeleland region to South Africa, transnational kinship relations and a long history of migration from the region to South Africa. The long history of migration has led to the emerge of networks between migrants and those who stay behind which facilitate migration. Parents, siblings, friends and neighbours already working in South Africa provides links through which those at home obtain information about their intended destinations, are assisted to get jobs and how the settle in their destinations (Maphosa, 2011).

Climate mobility is a form of human adaptation to the impacts of climate change (Sall et al., 2011; Musah-Surugu et al., 2017; Maharjan et al., 2021; Detges et al., 2022). Sall et al. (2011) state that in some areas of Senegal, migration has become a strategy to escape poverty that results from climate change. The remittances, migrants send back home help in diversifying community livelihoods from agriculture. Mobility provides:

> … an opportunity for migrants to generate funds and send money home. It is a key factor in adaptation to climate change as a strategy for survival and for diversifying incomes.
>
> (Sall et al., 2011: v)

Maharjan et al. (2018) concur that mobility may serve as an adaptation strategy or support the adaptation capacity of households in areas that are vulnerable to the negative effects of climate change. They assert that:

> Migration offers households an opportunity to diversify livelihoods and spread risks in ways that make their households less vulnerable to the impacts of global environmental change.
>
> (p. 88)

As mobility is a form of adaptation, financing mobility should be part of climate adaptation. Climate financing is important in supporting vulnerable communities to adapt to the risks of climate as well as those of migration. It must therefore be driven by the needs of those communities at risk. Consultation with vulnerable communities would ensure equity and inclusion (The White House, 2021).

The government of Zimbabwe's response to mobility from Matabeleland to South Africa has been characterised by nonchalance and at times outright dismissal. For example, in 2015 Robert Mugabe, the then President of Zimbabwe claimed that people the Matabeleland region who crossed the border into South Africa did so out their own choices. He stated that most Zimbabweans who "jump the border" to South Africa are unskilled Kalangas (a dominant ethnic group in Matabeleland) who survived on crime because they are not educated enough to get jobs. He stated that the Kalangas prefer working in South Africa to going to school.

The lack of political will to address the root causes of mobility from Matabeleland encourages the continued occurrence of irregular cross-border movements and the abuse of migrants in the country of destination. It is however clear that the impacts of climate change contribute to human mobility in Matabeleland. The need to mitigate the impacts of climate change is urgent. Ionesco (2019), for example, proposes various interventions to mitigate the impacts of climate change. These include investing in climate and environmental solutions so that people will not have to leave their homes in a forced way. The other solution is that states of origin should bear the primary responsibility for their citizens' protection even if indeed their countries have not been the main contributors to global warming. They should therefore apply human rights-based approaches for their citizens moving because of environmental or climatic drivers. A holistic approach to addressing the impacts of climate change require both mitigative and adaptive strategies. It should combine actions that reduce or prevent activities that lead to climate change while at the same time helping the vulnerable to cope with the impacts. It requires action from all stakeholders including government, climate financiers, human rights groups, and the vulnerable communities. This should be with the view to guaranteeing access, equity, diversity, participation, and human rights.

Conclusion

Climate change is now widely acknowledged as a growing global problem. Substantial scientific evidence on the impacts of climate change, such tropical cyclones, floods, droughts, and desertification is now available. Both policymakers and researchers are increasingly focusing their attention on the effects of climate and ways to mitigate and adapt to them. An area where research is still required is how climate change affects migration. While it is estimated that climate change will lead to the migration of millions of people in future years, knowledge on how climate change will actually affect migration is still anecdotal. This is because climate change often has an indirect impact on migration and is often one of a multiplicity of factors that lead people to leave their homes. As an indirect cause of migration, climate change is often overlooked as focus is directed at more immediate causes. Trying to establish a direct causal relationship between climate change and migration may be seen as simplistic climate determinism which strips migrants of agency. In exercising their agency, human beings can choose from many possible courses of action, migration being just one of them. However, the contribution of climate change to migration has to be investigated. The concept of climate mobility provides a useful framework of linking human migration to climate change without seeking to establish a causal relationship. To understand the dynamics between human mobility and climate change requires an awareness of the connectedness of environmental and socio-economic issues. Achieving environmental justice and guaranteeing that communities adapt with climate change, requires that the interventions to do so incorporate the principles of social justice.

Evidence shows that Zimbabwe is already experiencing the effects of climate change which cause, among other things droughts which have become frequent in recent years. Matabeleland region is probably the worst affected by climate change in the country because of its already low rainfall patterns and drought proneness. Many writers on migration from Matabeleland have largely overlooked the impact of climate on migration and emphasised economic and political factors. No doubt decades of economic and political marginalisation have contributed significantly to large-scale migration from Matabeleland to South Africa. The emphasis on economic factors in explaining migration from Matabeleland to South Africa contributes to how they are treated in their country of destination. This means they are not seen as forced migrants but as people who voluntarily migrate is search of a better life. That is why they are often received with hostility leading, among other things, violent acts of xenophobia by the locals. People who are forced to leave their places of origin by unbearable climatic conditions have to be treated like refugees. However, the general view of migrants from Zimbabwe as people seeking to maximise economic opportunities prevents them from being treated with the sympathy that is extended to refugees. In its efforts to mainstream environmental issues in its developmental plans, Zimbabwe needs to consider migration as adaptation to climate change and hence a developmental issue. Zimbabwean public officials, however, generally do not view migration, especially from Matabeleland as adaptation to climate, but an outcome of people's choices is search of greener pastures. It is therefore unlikely that the mainstreaming of climate change in development plans will consider the climate change–migration nexus.

References

ActionAid (2020). Climate change drives migration in conflict ridden Afghanistan. ActionAid International. https://actionaid.org/publications/2020/climate-change-drives-migration-in-conflict-ridden-afghanistan.

Agyeman, J., Schlosberg, D., Craven, L. and Matthews, C. (2016). Trends and directions in environmental justice: From inequality to everyday life, community and just sustainabilities. *Annual Review of Environment and Resources, 41*, 321–340.

Akokpari, J.A. (2000). Globalisation and migration in Africa. *African Sociological Review, 4* (2), 72–92.

Alexander, J., McGregor, J. and Ranger, T. (2000). *Violence and Memory: One Hundred Years in the "Dark Forest" of Matabeleland, Zimbabwe*. James Currey.

Bezu, S., Demissie, T., Abebaw, D., Mungai, C., Samuel, S., Radeny, M., Huyer, S. and Solomon, D. 2020. Climate change, agriculture and international migration nexus: African youth perspective. CGIAR Research Program on Climate Change, Agriculture and Food Security CCAFS), Working Paper Series #324.

Bhatasara, S. and Nyamwanza, A.M. (2022). Climate injustice and the role of climate justice movements in Africa: The case of Zimbabwe's radicalisms and conservatisms. In E. Etieyibo, O. Katsaura and M. Musemwa (eds.). *Africa's Radicalisms and Conservatisms*, 187–209. Brill.

Black, B., Bennett, S.R.G., Thomas, S.M. and Beddington, J.P. (2011). Climate change: Migration as adaptation. *Nature, 478* (7370), 447–449.

Bloch, A. (2008). Zimbabweans in Britain: Transnational activities and capabilities. *Journal of Ethnic and Migration Studies, 34* (2), 287–305.

Boas, I., Wiegel, H., Farbotko, C., Warner, J. and Sheller, M. (2022). Climate mobilities: Migration im/mobilities and mobility regimes in a changing climate. *Journal of Ethnic Studies, 48* (14), 3365–3379.

Brazier, A. (2015). *Climate Change in Zimbabwe: Facts for Planners and Decision Makers*. Konrad-Edenauer Stiftung.

Brittell, C.B. (1993). *When They Read What We Write. The Politics of Ethnography*. Bergin and Curey.

Brown, D., Chanakira, R.R., Chatiza K., Dhliwayo M., Dodman, D., Masiiwa, M., Muchadenyika, D., Mugabe, P., and Zvigadza, S. (2012). Climate change impacts, vulnerability and adaptation in Zimbabwe. IIED Climate Change Working Paper # 3. https://www.researchgate.net/publication/364757122_Climate_change_impacts_vulnerability_and_adaptation_in_Zimbabwe_Climate_Change_IIED_Climate_Change_Working_Paper_Series_Climate_change_impacts_vulnerability_and_adaptation_in_Zimbabwe

Brown, O. (2008). Migration and Climate Change. IOM Research Series Paper #31.

Chanza, N. and Gundu-Jakarasi, V. (2020). Deciphering the climate change conundrum in Zimbabwe: An exposition. In J.P. Tiefenbacher (ed.) *Global Warming and Climate Change*. IntechOpen. https://doi.org/10.5772/intechopen.84934.

Chikodzi, D., Murwendo, T. and Simba, F.M. (2013). Climate change and variability in South-eastern Zimbabwe: Scenarios and societal opportunities. *American Journal of Climate Change, 2* (3a). https://doi.org/10.4236/ajcc.2013.23A004.

Chingarande, S.D. and Maphosa, F. (2007). Migration as a coping strategy: The case of illegal migrants from Zimbabwe to the United Kingdom and South Africa. In F. Maphosa, K. Kujinga and S.D. Chingarande (eds.), *Zimbabwe's Development Experiences since 1980: Challenges and Prospects for the Future*. OSSREA.

Chingarande, S.D., Mugano, G., Chagwiza, G. and Hungwe, M. (2020). Zimbabwe market study: Matabeleland North Province report. USAID Research Technical Assistance Centre, Washington, DC.

Concern Worldwide (2022). Climate justice explained. https://www.concern.net/news/climate-justice-explained.

Crush, J., Perbedy, S. and Williams, V. (2006). International migration and good governance in the Southern African Region. SAMP Policy Brief No. 17.

Crush, J., Tawodzera, G., Chikanda, A., and Tevera, D. (2017). *Living with xenophobia: Zimbabwean informal enterprise in South Africa* (rep. i-33). Waterloo, ON: Southern African Migration Programme. SAMP Migration Policy Series No. 77.

Crush, J., and Tevera, D. (2010). *Zimbabwe's Exodus: Crisis, Migration, Survival.* International Development Research Centre.

Crush, J., and Tawodzera, G. (2016). Migration and Food Security: Zimbabwean Migrants in Urban South Africa. Urban Food Security Series No. 23, Kingston, ON and Cape Town: African Food Security Urban Network, 1–54.

Darbon, D. (1992). Fluctuat nec mergitur: Keeping afloat. In S. Baymanham (ed.) *Zimbabwe in Transition.* Almqvist and Wiksell International, 1–23.

Detges, A., Wright, E. and Bernstein, T. (2022). A conceptual model of climate change and human mobility interactions. HABITATE Research Paper. Aldephi.

Downey, A. (2013). Exemplary subjects: Camps and the politics of representation. In T. Frost (ed.), *Giorgio Agamben, Legal, Political and Philosophical Perspectives.* Routledge, 119–142.

Farbotko, C., Thornton, F., Mayshofer, M. and Herman, E. (2022). Climate mobilities, rights and justice: Complexities and particularities. *Frontiers in Climate, 4,* 1026486. https://doi.org/10.3389/faclim.2022.1026486.

Frazer, N. (1998). Social justice in the age of identity politics: Redistribution, recognition and participation. WZB Discussion Paper #FS1 98–101, Berlin.

Frischen, T., Meza, I., Rupp, D., Wietler, K. and Hagenlocher, M. (2020). Drought risk to agricultural systems in Zimbabwe: A spatial analysis of hazard exposure and vulnerability. *Sustainability, 12* (3), 752. https://doi.org/10.3390/su12030752.

Gaidzanwa, R.B. (1999). Voting with their feet: Migrant Zimbabwean nurses and doctors in the era of structural adjustment. Nordiska Aftrikainstitutet, Research Report No. 11.

Gonese, F.T., Marongwe, N., Mukora, C. and Kinsey, B. (2002). Land reform and resettlement in Zimbabwe: An overview of the programme against selected international experiences. https://minds.wisconsin.edu/bistream/handle/1793/23060/LRRPOverview.pdf.

Gosling, A., Thornton, P., Chevallier, R. and Chesterman, S. (2020). Agriculture in the SADC region under climate change. SADC Information Briefing. https://releifweb.int/report/angola/agriculture-sadc.region-under-climate-change.

Hobane, A.P. (1999). The commercialization of gonimbrasia belina in Bulilima-Mangwe District: Problems and prospects. MPhil thesis submitted to the Center for Applied Social Sciences (CASS), University of Zimbabwe, Harare.

Hoffman, R. (2022). Contextualising climate change impacts on human mobility in African drylands. *Earth's Future, 10* (6). https://doi.org/10.1029/2021EF002591.

Human Right Watch (2005). "Clear the filth": Mass evictions and demolitions in Zimbabwe: Background Briefing. The implementation of Operation Murambatsvina, September, 2005. https://www.hrw.org/legacy/background/africa/zimbabwe0905/4.htm.

Hunter, R., Crespo, O., Coldrey, K., Cronin, K. and New, M. (2020). Climate change and future crop suitability in Zimbabwe. University of Cape Town, South Africa. Research Highlights, undertaken in support of Adaptation for Smallholder Agriculture Programme ASAP, Phase 2. International Agricultural Fund for Agricultural Development (IFAD), Rome.

IOM (2014). IOM outlook on migration, environment and climate change. https://publications.iom.int/system/files/pdf/mecc_outlook.pdf.

IOM (Egypt Office) 2020. *Migration and Climate Change Nexus.* IOM.

Ionesco, D. (2019). Let's Talk About Climate Migrants, Not Climate Refugees. United Nations Sustainable Development. https://www.un.org/sustainabledevelopment/blog/2019/06/lets-talk-about-climate-migrants-not-climate-refugees/.

Machinya, J. (2019). Undocumented Zimbabweans in South Africa: Working in constant fear of arrest and deportation. *Labour, Capital and Society, 49* (2), 91–114.

Madebwe, C. and Madebwe, V. (2017). Contextual background to the rapid increase in migration from Zimbabwe since 1990. *Inkanyiso: Journal of Humanities and Social Sciences, 9* (1), 27–36.

Maharjan, A, Hussain, A, Bhadwal, S, Ishaq, S, Saeed, B.A, Sachdeva, I, Ahmed, B, Hussain, S.M.T, Tuladhar, S and Ferdous, J (2018). Migration in the Lives of Environmentally Vulnerable Populations in Four River Basins of the Hindu Kush Himalayan Region. http://hi-aware.org/wp-content/uploads/2018/10/working-paper-20.pdf.

Maphosa, F. (2007). Remittances and development: The impact of migration to South Africa on rural livelihoods in Southern Zimbabwe. *Development Southern Africa, 24* (1), 123–136.

Maphosa, F. (2011). *Multiple Involvements or Multiple Exclusions? Transnational Experiences of Communities on the Zimbabwe-South Africa Borderlands.* OSSREA.

Maphosa, F. and Ntau, C. (2020). Undocumented migrants as *homo sacer.* Cases from Botswana and South Africa. *Journal of Asian and African Studies, 56* (4), 872–888.

Maviza, G. (2020). Transnational migration and families, continuities and changes along processes of sustained migration: A case of Tsholotsho in Matabeleland North, Zimbabwe. PhD thesis submitted to the University of the Witwatersrand, South Africa.

Maviza, G., Maphosa, M., Tshuma, N., Dube, Z. and Dube, T. (2019). Migrant remittances inspired enterprises in Tsholotso: Issues of sustainability. *African Human Mobility Review, 5* (1), 1459–1481.

Mbiba, B (2005). Zimbabwe's global citizens in "Harare North": Overview and implications for development. Peri-Net Working Paper No. 14.

Mbiyozo, A.N. and le Roux. (2021). Climate finance isn't reaching Southern Africa's most vulnerable. Institute for Security Studies. https://issafrica.org/iss-today/climate-finance-isnt-reaching-southern-africas-most-vulneable.

McDonald, D.A., Zinyama, L, Gay, J de Vletter, F and Mattes, R (2000) Guess who's coming to dinner: Migration from Lesotho, Mozambique and Zimbabwe to South Africa. *International Migration Review, 34* (3), 813–841.

Mollenkamp, D.T. (2022). Social justice: Meaning and main principles explained. *Investopedia.* https://www.investoedia.com.terms/s/social-justice.asp.

Moyo, K. (2021). South Africa reckons with its status as a top immigration destination: Apartheid history and economic challenges. Migration Policy Institute. https://www.migrationpolicy.org/articles-south-africa-immigration-destination-histories.

Mugandani, R., Wuta, M., Makarau, A. and Chipindu, B. (2012). Reclassification of agroecological regions of Zimbabwe in conformity with climate variability and change. *Africa Crop Science Journal, 20* (2), 361–369.

Munyoka, E. (2020). Causes of irregular migration of people from Zimbabwe to South Africa in the post Mugabe regime. *African Journal of Education and Social Science, 7* (3), 34–46.

Musah-Surugu, I.J., Ahenkan, A. and Bawole, J.N. (2017). Migrants' remittances: A complimentary source financing adaptation to climate change at the local level in Ghana. *International Journal of Climate Change and Management, 10* (1), 178–196.

Musemwa, M (2006). Disciplining a dissident city: Hydropolitics in the city of Bulawayo, Matabeleland, Zimbabwe, 1980-1994. *Journal of Southern African Studies, 32* (2), 238–254.

Mushore, T.D., Mhizha, T., Manjowe, M., Mashawi, L., Matandirotya, E., Mushonjowa, E., Mutasa, C., Gwenzi, J. and Mushambi, G.T. (2021). Climate change adaptation and migration strategies for small-holder farmers: A case of Nyanga District in Zimbabwe. *Frontiers in Climate, 3,* 676495. https://doi.org/103389/fclim.2021.676495.

Muzerengi, T. and Tirivangasi, H.M. (2019). Small grain production as an adaptive strategy to climate change in Mangwe District, Matabeleland South in Zimbabwe. *Jamba, Journal of Disaster Risk Studies, 11* (1), A652. https://doi.org/10.4102/Jamba.v11i1/652.

Myers, N. (2005). Environmental refugees: An emergent security issue. Paper presented at the 13th Economic Forum in Prague, May, 22–23.

Naser, M.M. (2021). *The Emerging Global Consensus on Climate Change and Human Mobility.* Routledge.

Ndlhovu, N.G. (2021). "Moment of Madness": Ethnic marginalisation and exclusionary development: Are they not social development issues? *Southern African Journal of Social Work and Social Development, 33* (2). Gale Academic OneFile, link.gale.com/apps/pub/713W/AONE?u=anon-d98aefe5&sid=bookmark-AONE.

Ndlovu, E., Prinsloo, B. and le Roux, T. (2020). Impact of climate change an variability on traditional farming systems: Farmers' perceptions from south west semi-arid Zimbabwe. *Jamba: Journal of Disaster Risk Studies, 12* (1), a 472. https://doi.org/10.4102/jamba.v12i1:742.

Nicholson, C.T.M. (2011). Is the "environmental migration" nexus and analytically meaningful subject for research. Centre for Migration, Citizenship and Development, University of Swansea, Working Paper Series #104.

Nzima, D., Moyo, P. and Duma, V. (2016). Theorising migration-development interactions: Towards an integrated approach. *Migration and Development.* https://doi.org/10.1080/21632324.2016.1147897.

Ornstein, A.C. (2017). Social justice: History, purpose and meaning. *Social Science and Public Policy, 54,* 541–548.

Oucho, J. (2007). Migration is Southern Africa: Migration management initiatives for SADC members states. Institute for Security Studies Paper 157.

Paris Agreement. (2015). UN DOC.FCCC/CP/2015/10/Add.1 decision 1/CP.21

Pasura, D. (2010). Competing meanings of the diaspora: The case of Zimbabweans in Britain. *Journal of Ethnic and Migration Studies, 36* (9), 1445–1461.

Piguet, E. (2011). The migration/climate change nexus: An assessment. Paper presented at the International Conference on Rethinking Migration: Climate, Resource Conflicts and Migration in Europe, 13–14 October, 2011.

Podesta, J. (2019). The climate crisis, migration and refugees. Paper prepared for the 2019 Brookings Blum Roundtable. https://www.brookings.edu/wp-content/uploads/2019/07/Brookings_Blum_2019_climate.pdf.

Portner, H. O., Roberts, D. C., Adams, H., Adler, C., Aldunce, P., Ali, E., Ara Begum, R., Betts, R., Bezner Kerr, R., Biesbroek, R., Birkmann, J., Bowen, K., Castellanos, E., Cissé, G., Constable, A., Cramer, W., Dodman, D., Eriksen, S. H., Fischlin, A., … Zaiton Ibrahim, Z. (2022). Climate Change 2022: Impacts, Adaptation and Vulnerability. IPCC. https://edepot.wur.nl/565644

Ravenstein, E. (1889). The laws of migration. *Journal of the Statistical Society of London, 48* (2), 167–235.

Rawls, J. (1971). *A Theory of Justice.* Harvard University Press.

Roth, M.R. and Bruce, J.W. (1994). Land tenure, agrarian structure and comparative land use efficiency in Zimbabwe: Options for land tenure reform and land redistribution. University of Wisconsin-Madison, Land Tenure Centre Research Paper 117.

Sachikonye, L.M. (1988). Rethinking about labour markets and migration in Southern Africa, *Southern Africa Political and Economic Monthly, 11* (4):11–20.

Sherbinin, S. (2020). Climate impacts as drivers of migration. Migration Policy Institute. https://migrationpolicy.org/article/climate-impacts-drivers-migration.

Shimeles, A. (2010). Migration patterns, trends and policy issues in Africa. African Development Bank Group, Working Paper No. 19.

Stephens, C. (2007). Environmental justice: A critical issues for all environmental scientists everywhere. *Environmental Research Letters, 2*. https://doi.org/10.1088/1748–9326/2/4/045001.

Stojanov, R., Rosengaertner, S., de Sherbinin, A. and Nawrotzki, R. (2021). Climate mobility and development. *Population and Environment, 43*, 209–231.

Szaboova, L. and Colon, C. (2020). Concepts, contexts and categorisations of climate mobility. UNICEF, Climate Mobility and Children: A Virtual Symposium, 3–4 November, 2020.

Tacoli, C. (2009). Crisis or adaptation? Migration and climate change in a context of high mobility. *Environment and Urbanisation, 21* (2), 513–525.

Tadesse, D. (2010). The Impact of Climate Change in Africa. Institute for Security Studies Paper 220. https://www.files.ethz.ch/isn/136704/PAPER220.pdf.

The White House. (2021). Report of the impact of climate change on migration. The White House.

Thebe, V. (2011). From South Africa with love: The quest for livelihood reconstruction in South-Western Zimbabwe. *Journal of Modern African Studies, 49* (4), 647–670.

Tshuma, M. (2020). Matabeleland cattle deaths worry stakeholders. CITE. https://cite.org.zw/matabeleland-cattle-deaths-worry-stakeholders.

United Nations. (1994). United Convention to Combat Desertification (UNCCD). https://cil.nus.edu.sg/databasecil/1994-united-nations-convention-to-combat-desertification-in-those-countries-experiencing-serious-drought-and-or-desertification-particularly-in-africa/

United Nations. (2006). Social justice in an open world: The role of the United Nations. United Nations.

United Nations (2014). *United Nations Convention to Combat Dessertification in those Countries Experiencing Serious Drought and/Dessertification, Particularly in Africa.* https://catalogue.unccd.int/936_UNCCD_Convention_ENG.pdf

United Nations. (2015). Framework convention on climate change. UNDOC.FCCC/CP/2015/h.9/Rev/1 (December 12, 2015).

United Nations. (2017). UN Climate Change Annual Report 2017. https://unfccc.int/resource/annualreport.

United Nations. (2019). Migration and the Climate Crisis: The UN's Search for Solutions. https://news.un.org/en/story/2019/07/1043552

United Nations. (2020). Climate change is an increasing threat to Africa. https://unfccc.int/news/climate-change-is-an-increasing-threat-to-africa

United Nations. (2022). *Southern Africa: A Regional Response to the Smuggling of Migrants.* United Nations.

United Nations Development Programme. (2021). Climate Change Adaptation: Zimbabwe. https://www.daptation-undp.org/explore.eastern-africa/Zimbabwe.

United Nations Higher Commission for Refugees. (2020). Midyear trends 2019. https:www.unhcr.org/statistics/unhcrstats/5e57doc57/mid-year-trends-2019.

Van Onseln, C. (1976). Chibaro: African Mine Labour in Southern Rhodesia, 1900–1933. Pluto Press.

Vinke, K., Rottmann, S., Gornott, C., Zabre, P., Schwerdtle, N. and Sauerborn, R. (2022). Is migration an effective adaptation to climate-related agricultural distress in Sub-Saharan Africa? *Population and Environment, 43*, 319–345.

Waldinger, M. and Fankhauser, S. (2015). Climate change and migration in developing countries: Evidence and implications for PRISE countries. Centre for Climate Change Economics and Policy, Policy Paper.

Wamukonya, N. and Rukuto, H. (2001). Climate change implications for Southern Africa: A gendered perspective. https://www.energia.org/assets/2015/06-climate-change-implications-for-Southern-Africa,pdf.

Warner, K. (2010). Global environmental change and migration: Governance challenges. *Global Environmental Change, 20* (3), 402–413.

World Bank (2018). Mixed migration, forced displacement and job outcomes in South Africa. World Bank.

World Food Programme (WFP) (2021). Climate change in Southern Africa: A position paper for the World Food Programme in the Region. https://executiveboard.wfp.org/document_download/WFP-0000129015.

Wulff, M. and Newton, P. (1996). Mobility and change: Australia in the 1990s. In P. Newton and M. Bell (eds.) *Population Shift: Mobility and Change in Australia.* Australian Government Publishing Service (AGPS), pp. 426–443.

Young, T., Tucker, T., Galloway, M., Manyike, P., Chapman, A. and Myers, J. (2010). Climate change and health in the SADC region: A review of the current state of knowledge. SADC Climate Change and Health Synthesis Report. https://open.uct.ac.za/bitstream/item/2144/climate_change_health.pdf?sequence=1.

Zanamwe, L. and Devillard (2009). Migration in Zimbabwe: A country profile 2009. A study commissioned by ZIMSTAT/IOM/EU. https://publications.iom.int/systems/files/pdf/mp_zimbabwe.pdf.

Zinyama, L.M. (1990). International migration to from Zimbabwe and the influence of political changes on population movements, 1965–1987. *International Migration Review, 24* (4), 748-767.

13 Climate Change and Gendered Migration Patterns in Southern Africa

A Review

Divane Nzima and Gracsious Maviza

Introduction

Climate change is among the most topical concerns that the world is presently grappling with. Efforts to address this challenge have been largely informed by the principles of a just transition and the need for social equity that ensures people are at the core of the initiatives and no one is left behind (Allwood, 2020). The need for social justice and equity stems from the realisation that the risks and effects associated with climate change continue to disproportionately affect those least responsible and least able to adapt (Chang et al., 2014). Considering this, responses to the effects of climate change must ensure social justice and equity. There are two primary forms of social justice – distributive and procedural. Distributive social justice refers to how resources, benefits, and burdens are allocated across different population contingents. Procedural social justice is concerned with "the fairness and transparency of the processes used to make decisions about societal goals, i.e., 'who decides' and 'who participates' in decision-making processes" (Preston et al., 2014, p. 14). Within this framework, climate justice should thus involve the full and equitable participation of all in making decisions about climate change response strategies despite their social class, gender, race, and geographical location. In this chapter, we submit that embedding a gender lens has been one of the most valuable initiatives in the drive towards climate justice, emphasising social equity in climate change response strategies. Consistent with the social equity framework, we focus on how the persistent challenge of climate change continues to influence gendered migration patterns in Southern Africa and whether efforts to address this have upheld the principles of social justice. Although this is a critical area of focus, in Africa, there are still very limited studies focusing on this nexus. Ongoing debates still question whether there is a link between climate change and migration, let alone gender. Undeniably, climate change is not gender neutral (Masinga et al., 2021). It has differential effects on women and men due to the differences in responsibilities, access to resources and assets, and decision-making power, which significantly compounds inequalities and vulnerabilities (Chingarande et al., 2020). Although this is acknowledged, there is still very limited data to show linkages and make concrete conclusions on the nexus between climate change, migration, and gender. Notwithstanding, the subject remains critical in attempts to establish the

DOI: 10.4324/9781003397120-16

linkages therein and expose the inherent gaps in the study of climate change, climate justice, migration, and gender, hence this chapter.

Lim Kam Sian et al. (2021) argue that there is a pressing need to gather knowledge about trends in climate change to plan for and make informed decisions about sustainable development. Southern Africa has unique problems that include a bulging population and general poor economic performance, making its susceptibility to climate change and variability more pronounced (Lim Kam Sian et al., 2021). Nhamo et al. (2019) argued that climate change would have an impact on the future and current farming in Southern Africa and the rest of the continent. In addition, this will affect food systems as climate variability modifies environments substantially, owing to shifts in seasons (Nhamo et al., 2019). In Southern Africa, the agricultural sector continues to be negatively affected by climate change which inevitably affects the region's capacity to meet its food demands (Nhamo et al., 2019; Lim Kam Sian et al., 2021). Considering its growing population, Southern Africa faces water deficits due to the increased intensity and frequency of droughts. As a result, many countries in Southern Africa fail to meet their food requirements because they largely depend on rain-fed agriculture for food production. Where irrigation is possible, water scarcity remains a lingering problem as the agricultural sector must compete with other economic sectors for the limited water resources (Nhamo et al., 2019; Lim Kam Sian et al., 2021). In all this, gender takes centre stage because societal structural inequalities make women more susceptible and vulnerable to climate change (Allwood, 2020). With the increase in droughts and water shortages and food insecurity, women in most African communities bear the brunt of care responsibilities to ensure households have water and food. As a result, inequalities become proliferate, resulting in compromised social equity.

Studies have shown that increasing temperatures, changing rainfall patterns, and rising sea levels, as well as land and water degradation, continue to hinder attempts at managing problems driven by increasing demand for water and food (Brown & Funk, 2008; Camill, 2010; Nhamo et al., 2019). Therefore, this calls for innovative ways of dealing with the problem, such as new methods of employing technology to build resilience and adaptation. While calling for such initiatives, it is critical to ensure that the efforts are gender inclusive to ensure social equity. These initiatives should go beyond the normative mark of gender inclusion marked by a mere numerical presence of women in climate decision-making. Rather, they should ensure "inclusion at all levels of decision making [to guarantee inclusion of] gender equality and gender justice objectives in policies, action plans and other measures" (Allwood, 2020, p. 8). By so doing, these initiatives will ensure social equity at different levels of dealing with climate change to guarantee no one is left behind. Moreover, if well implemented, social equity principles will inform strategies that empower communities to own their climate adaptation and resilience strategies and become active players in the broader climate change discourses that shape their lived realities. These will directly address their sociocultural realities to ensure an inclusive and more socially equitable approach.

Mason (2001) argued that Southern Africa has been prone to El Nino events. He attributes this to the below-normal rainfall over much of Southern Africa (Mason,

2001). Studies that investigated the changes in precipitation variability and trends during the 20th and 21st centuries using rain gauge data in Southern Africa suggest that precipitation shows considerable spatial variability, unlike temperature, whose variation is uniform (Jury, 2013; Kusangaya et al., 2014; Lim Kham Sian et al., 2021; Thoithi et al., 2021). According to Thoithi et al. (2021), Southern Africa is prone to recurrent floods and droughts that adversely affect the growing rural population, who are largely economically disadvantaged. Thoithi et al. (2021) suggest that the highly variable rainfall across Southern Africa disproportionately affects the poor rural population that is heavily dependent on rain-fed agriculture for food production.

Southern Africa has experienced several climate catastrophes that have left hundreds of people dead, and thousands displaced. In April 2022, the city of Durban and surrounding areas in South Africa were hit by devastating floods and mudslides. This climate disaster left over 400 people dead, thousands more displaced, and property worth millions destroyed. This came after South Africa was still recovering from the Durban floods that occurred in April 2019. In the year 2022 alone, from January to March, Madagascar was hit by five tropical storms and cyclones. These included Ana, Batsirai, Dumako, Emnati, and Gombo. All this caused immense damage and loss of life and left thousands of people displaced. Mozambique has also been a regular victim of tropical storms and cyclones as evidenced by the devastating effects of Idai in 2019, Eloise in 2021 and Ana in 2022 which left thousands of people from Beira and surrounding areas displaced. Most of the tropical storms and cyclones that affect Madagascar usually end up affecting Mozambique and some parts of the Eastern Highlands in Zimbabwe. Some examples of these tropical storms and cyclones that have moved inland and affected the Eastern Highlands in Zimbabwe include Japhet in 2003, Idai in 2019, Chalene in 2020, Eloise in 2021 and Ana in 2022.

In response to the destruction of property, livelihoods, and mass displacement from homes, which disproportionately affect women, surviving victims of these natural disasters caused by climate change seek alternative livelihood strategies. Migration becomes one of the strategies of choice as it allows people to move to safer areas and also enables them to diversify their livelihoods. Previous studies on migration in Africa have shown that loss of livelihoods due to ecological factors as well as unabated poverty that manifest in high levels of unemployment are some of the key drivers of eternal and international migration (Maphosa, 2007; Abu et al., 2014; Strobl & Valfort, 2015; Mastrorillo et al., 2016). In Southern Africa, limited studies have assessed the correlations between climate change and migration. One study that has tried to link climate change as a driver of migration was done by Mastrorillo et al. (2016), where secondary data were used to assess the impact of climate variability on internal migration flows in post-apartheid South Africa. The results of this study showed that an increase in positive temperature extremes and positive and negative excess rainfall at the origin act as a push factor and enhance out-migration, mostly among poor rural black communities (Mastrorillo et al., 2016). This is an important finding because the effects of climate change in Southern Africa are often heavily felt by poor rural communities whose livelihoods are centred on subsistence farming. The role of climate change in influencing migration is sometimes downplayed or poorly characterised because its role as a contributing factor can sometimes be disguised in other common migration push factors such as joblessness and lack of economic opportunities (Abu et al., 2014; Strobl & Valfort, 2015). Therefore, there is a need for more studies that invariably seek to find a significant causal relationship between the effects of climate variability and both internal and international migration in Southern Africa.

Southern Africa has a well-documented history of migration that was driven by the migrant labour system that was centred around the discovery of gold in South Africa in the 19th century (Nzima & Moyo, 2017). South Africa has always been the major recipient of migrants in the region, with countries such as Malawi, Mozambique, Lesotho, and the former Transkei homeland sending men to work in the mines of Johannesburg (Nzima & Moyo, 2017). During that time migration was a preserve for males as most jobs were mostly suitable for men. As a result, women were often left to look after their children and attend to subsistence farming in rural areas. In addition, some of the migration flows within the region were driven by conflict during the liberation wars in the 1960s, while many other flows happened when there were incidences of post-colonial unrest in some of the newly independent state in Southern Africa in the 1980s (Zinyama, 1990; Nzima & Moyo, 2017). Until today, South Africa continues to be the main recipient of migrants from across the region and beyond as many post-colonial African countries face continued political and economic turmoil (Nzima & Moyo, 2017). Some migrants flee adverse economic conditions that can be attributed to the effects of climate change that have destroyed livelihoods and left many people displaced. As such, there is both a mixture of males and females in the migrants' stocks. Therefore, this paper is timely as it seeks to review the effects of climate change and how these have influenced gendered migration patterns in Southern Africa. In doing so, we look at climate change and agro-based livelihoods. We also review the culture of migration, the feminisation of migration in the face of increasing climate variability and migration as a means of adaptation and resilience in the face of climate variability.

Gender, Climate Change, and Agro-Based Livelihoods

Although living in the same families and communities, women and men live their private and public lives in different places and inevitably occupy different cultural spaces in society (Nagel, 2015). These relative places and spaces in society perpetuate socially constructed differences between the two groups assigned on the binaries of gender – masculine or feminine. According to Carling (2005), gender refers to the cultural valuing and interpretation of a person's biology resulting in locally accepted ideologies of what it is to be a woman or man. Similarly, other scholars have also conceptualised gender as a manifestation of dynamic and context-specific social constructs that assign different roles, expectations, and positions used to identify men and women in societies (Nagel, 2015; Pearse, 2017). These conceptions of gender form the basis of gender relations, wherein different roles are assigned to individuals based on whether they are male or female (Radel

& Schmook, 2009). Thus, gender relations are a social construct assigned according to one's biological makeup designed to uphold social order (Pinnawala, 2009). However, in the process, these have perpetuated power hierarchies, resulting in significant inequalities between men and women (Jayachandran, 2015). This is especially true for most African societies where there is notable patriarchal dominance, with male members as heads of families having legitimate power and authority to control others within the family unit (Mangleburg et al., 1999; Mashiri & Mawire, 2013; Helman & Ratele, 2016). In this context, the duties and responsibilities of different members are framed by this gender ideology and are embedded in notions of masculinity for men and femininity for women (Masanja, 2012).

Ideally, masculinity for men entails breadwinning responsibilities related to providing for the family, while femininity for women is highly linked to nurturing and caring practices (Donaldson & Howson, 2009; Masanja, 2012). This was the case even in pre-historic hunter-gathering communities, where women were the gatherers while the men were the hunters, a practice which meant women provided for daily sustenance and their livelihood activities heavily leaned on nature (Dankelman, 2010). During the colonial era, there was an introduction of the cash economy which changed most of these practices, for example, in rural Southern African communities (Potts & Mutambirwa, 1990; Donaldson & Howson, 2009; Hunter, 2010). Due to the introduction of the cash economy, most men in colonial Africa moved from their rural homes in search of employment opportunities in urban centres (Potts & Mutambirwa, 1990; Posel, 2003). This labour regime promoted the contractual employment of males while women remained in the communal lands tending to agricultural production to ensure their families were well-fed (Potts & Mutambirwa, 1990; Hunter, 2010). As a result, while the men were out working for wages, women remained at home, engaging in invisible and often undervalued and unpaid care work (Adisa et al., 2019; Maunganidze, 2020). Women took on reproductive, productive, and community roles, what Moser (2012) refers to as the triple roles of women in society, disproportionately loading them with the burden of ensuring families are sustained on a daily basis.

Owing to the colonial legacy, similar trends are observed in the contemporary landscape where women still disproportionately dominate rural areas in most African countries (Patel, 2020). For example, in Sub-Saharan Africa, including the Southern African Development Community (SADC) region, women are the backbone of rural communities.[1] A survey covering 34 African countries revealed that 55% of the women in these countries live in rural areas, with the least being Gabon (21%) and Eswatini the most at 83% (Patel, 2020). In the SADC region, with an estimated population of 300 million, 51% are women and 60% of the total female population resides in rural areas.[2] This is further corroborated by statistics showing that in Malawi, 81% of the women live in rural areas while Zimbabwe and Mozambique have 63% and Zambia 56%.

Considering the foregoing and given that most societies in Southern Africa heavily depend on rain-fed subsistence agriculture for livelihoods, and women take on the responsibility of agricultural production to provide for their families, climate change would inevitably have gendered impacts on livelihoods. According to Chidakwa et al. (2020), climate change compounds gendered dimensions of vulnerability embedded in existing social inequalities and gendered division of labour. This is further complicated by women's heavy reliance on natural resources for different aspects of their livelihoods, i.e., food, income, and energy, their limited rights to productive resources and the gendered power asymmetries in societies (Garutsa et al., 2018). These factors compounded aggravate women's vulnerability and susceptibility to climate change and its stresses (Garutsa et al., 2018; Chidakwa et al., 2020).

Therefore, given the differences in gendered livelihood pursuits – where men chase after money and women ensure the daily sustenance of families – the effects of climate change are not uniformly felt across genders (Nagel, 2015; Pearse, 2017). The gendered vulnerabilities are more pronounced for women as they are the ones manning agricultural production while the men are away looking for money. Therefore, the effects of climate change, such as extreme climatic events and weather conditions – strong winds, floods, and droughts, among others – negatively impact crop production, adversely affecting agro-based livelihoods (Pearse, 2017; Garutsa et al., 2018). Moreover, given that women form the majority of rural dwellers in most Southern African communities and that most of these communities depend on agro-based livelihoods, climate change thus has disproportionate impacts on women (Pearse, 2017). These demonstrate what has come to be known as the concept of gendered vulnerability to climate change which acknowledges that impacts of climate change are not intrinsic characteristics for both genders, but instead, they are expressions of prevailing inequalities and power asymmetries between men and women across different societies in the world (Dankelman, 2010; Nagel, 2015; Pearse, 2017).

Given that women's livelihood pursuits are mostly agro-based or nature-based and heavily depend on the availability of reliable water, it can be concluded that climate change adversely impacts agro-based livelihoods (Nagel, 2015; Pearse, 2017; Chidakwa et al., 2020). Inevitably, women become the most affected due to the gendered division of roles and responsibilities within families and communities at large (Dube et al., 2016; Pearse, 2017; Garutsa et al., 2018; Chidakwa et al., 2020). For Pearse (2017), these gendered differences in terms of responsibilities and roles as well as positions in society actively influence the disparities in men's and women's vulnerabilities to climate change. Therefore, the gendered dimensions of climate change directly reflect the gendered aspects of social life in societies. More specifically, gendered inequalities within families and communities aggravate the unequal effects of climate change, disproportionately making women more vulnerable than men (Dankelman, 2010; Nagel, 2015; Pearse, 2017; Garutsa et al., 2018; Chidakwa et al., 2020). Although there is limited data to prove the causal relationship, these gendered differential impacts of climate change on agro-based livelihoods may be a potential cause of the growing incidence of the increase in female migration stocks in Southern Africa. As already discussed in the foregoing, women experience first-hand, the resultant increases in water shortage, crop failures due to droughts and non-viability of livestock production. Given that these are directly linked to issues of food availability and household subsistence, they may ultimately lead to the decision by women to migrate. Therefore, there is a need for robust studies to address the data gap on the foregoing so that concrete conclusions may be drawn on this critical subject from a social equity perspective.

The Culture of Migration

Migration is a common phenomenon in Southern Africa with a long-standing history (Murray, 1981; Yabiku et al., 2010; Posel & Marx, 2013). It has actively shaped economies for decades, with governments making efforts to control it with limited success (Potts & Mutambirwa, 1990; Ncube et al., 2014; Zack et al., 2019; Musoni, 2020). This migration has become one of the most dominant livelihood strategies for many in the region as a reaction to several socioeconomic, agro-ecological and political issues in the different countries of the region (Landau, 2011; Ncube et al., 2014; Maviza, 2020; Moyo, 2020a). Although international in some instances, much of the mobility has been within the region, with countries like South Africa, Botswana, and Namibia hosting the largest contingents of migrants from other countries in the region (Nyamunda, 2014; Moyo, 2020a). Much of this is attributed to the opportunities available in the economies because of their better standing in terms of economic growth and development (Moyo, 2020a). Notably, this regional mobility has been highly irregular, transmuting from individual illicit labour recruiting in pre-colonial times to post-independence assisted cross-border crossings, where syndicates made of state and non-state actors play a pivotal role (Tshabalala, 2019; Ndlovu & Landau, 2020). Although common and long-standing, the mobilities have taken different forms and characters through different epochs.

In the colonial times, due to the introduction of cash economies and the introduction of the tax system, internal migration increased as men emigrated from the rural areas to urban centres in search of wage employment (Potts & Mutambirwa, 1990; Hunter, 2010; Posel & Marx, 2013). Often, these labour regimes preferred employing single male migrants on a contract basis without permanent settlement in the cities (Potts & Mutambirwa, 1990; Hunter, 2010). Examples of these were noted in Zimbabwe (then Rhodesia) and South Africa, where men migrated to cities in droves to search for wage employment, leaving their families behind (Hunter, 2010; Potts, 2010). Regional mobility was also somewhat organised through the institutionalised contract labour system that provided labour for the mines in the 1800s (Mabhena, 2010; Mlambo, 2010; Moyo, 2017). Although much of it was formal, informal and often, clandestine mobility was also a common feature (Mlambo, 2010; Maviza, 2020). During the period, South Africa and Zimbabwe topped the destination list, attracting migrants to work in the mines and manufacturing industries, respectively (Wentzel, 2003). For South Africa, migrants were recruited under the administrative system known as the Witwatersrand Native Labour Association (WNLA). At the same time, for Zimbabwe (then Rhodesia), recruitment was done under the Rhodesia Native Labour Bureau (RNLB) labour recruitment agency (Mlambo, 2010). In both cases, the migrant workers came from either Nyasaland (now Malawi), Northern Rhodesia (now Zambia) or Mozambique (Scott, 1954; Mlambo, 2010; Maviza, 2020). Notably, both systems preferred men due to the nature of the jobs available, leaving women in the communal lands nurturing families and overseeing agricultural production (Potts & Mutambirwa, 1990).

After independence, migration remained a common feature in most countries of the region (Delius, 2017; Moyo, 2020a). Currently, although there have been significant changes to the character of migration, it remains an important characteristic of the region (Moyo, 2020a, 2020b). Unlike in the past, when migration was a preserve for men, women have also become active and independent migrants (Thebe, 2019; Thebe & Maviza, 2019; Matose et al., 2022). This has led to the theoretical concept of feminisation of migration driven by the changing character and nature of the jobs available in the destination countries (Yeates, 2004; Pillinger, 2007; Zack et al., 2019; Takaindisa, 2020). Furthermore, this has been facilitated by the advent of transnationalism enabled by developments in transport and telecommunication technologies which have made regional and international mobility more accessible (Baldassar & Merla, 2014; Maviza, 2020). In the Southern African region, just like in the colonial era, contemporary mobility continues to be driven by the ever-deteriorating economic conditions in some countries as well as political unrests and social affinities for some communities living in borderlands (Maphosa, 2010, 2012; Matose et al., 2022).

The Feminisation of Migration in the Face of Increasing Climate Variability

Submissions have been made that much of contemporary migration flows in Southern Africa are driven by failed economies and adverse political conditions (Maphosa, 2010; Matose et al., 2022). However, in this paper, we make an ambitious proposition by submitting that perhaps the new trend where there are now more females migrating within the region, what has been termed the feminisation of migration could also be attributed to the increasing effects of climate change. In making this submission, we are fully aware of the criticism that we might attract, however, given that effects of climate change can sometimes be hidden in other migration push factors (see Abu et al. 2014; Strobl & Valfort, 2015) such as lack of employment, limited livelihood opportunities and civil unrest that may be as a result of displacement, this is an important link to further explore. Therefore, while being over-ambitious in our submission we also forward this as an invitation for more targeted research to explore the link between the increasing effects of climate change and the surge in the numbers of female migrants in Southern Africa.

According to Gouws (2010), despite an increase in the number of women migrating in Africa, there are still limited studies that focus on the migration of females. There are now more women who are migrating independently as migration in Africa has become a way of life owing to depleting livelihood options (Gouws, 2010). Migration has traditionally been a preserve for males, where males migrated to the cities in search of long-term employment. Gouws (2010) argues that women migrate for different reasons as trends suggest. They migrate for short periods of time to participate in cross-border trading and sometimes to look for seasonal employment. This cross-border trading is a common feature among Zimbabwean women who cross to South Africa, Botswana, Mozambique and Zambia to trade their wares. In addition, women from Lesotho often migrate to look for seasonal agricultural work on the farms of the Free State in South Africa.

There is also growing evidence that the gender patterns of internal migration stocks in South Africa are now dominated by women (Camlin et al., 2014). Camlin et al. (2014) argue that while female migration was primarily linked to marriage, South Africa is experiencing a feminisation of internal migration driven by the increasing labour market participation among women. In their study in KwaZulu Natal, Camlin et al. (2014, p. 2) found that "women were somewhat more likely than men to undertake any migration, but sex differences in migration trends differed by migration flow, with women more likely to migrate into the area than men and men more likely to out-migrate." The reason for this change in the gendered migration patterns is a new phenomenon that needs further scrutiny. Gouws (2010) suggested that as men lose jobs in the mines women are pressed to migrate for trade and seasonal work. We argue that some of the reasons for this increase in the number of females migrating within Southern Africa are linked to the increasing effects of climate change that have compromised the efficacy and sustainability of rural livelihoods. Moreover, with the increase in failed migration by men and the unproductivity of land – due to climate change – women have transitioned into the male breadwinner role by engaging in migration to get paid work and support their children (Maviza, 2020). Gouws acknowledges that food insecurity could be one of the factors that are pushing women to migrate for the purpose of doing trade and seasonal work. We submit that most of the food insecurity in Southern Africa can be linked to increasing effects of climate change such as prolonged droughts, water scarcity, floods, and rising temperatures that have a direct adverse effect on subsistence farming. Given the foregoing, there is a need for more targeted primary research on gender, climate change and migration patterns in Southern Africa. Moreover, women's movement has other far-reaching effects on families. Although migration becomes a strategy to address the effects of climate change, it may limit social equity within families, especially for the children left behind. The children's patterns of residence and authority may change as they move to be cared for by other relatives, this may also have detrimental effects on their psychological and emotional wellbeing (Maviza, 2020). Therefore, there is a need for more academic rigour in assessing how women's mobility in response to climate-induced push factors may foster or limit social equity within families, especially for the children left behind.

Migration as a Means of Adaptation to Climate Variability

Human mobility is an age-long phenomenon. It "involves the geographical movement of people from their area of origin to take up temporary or permanent residence in another area" (Kendall, 2012, p. 558). This mobility often happens on three distinct levels: international, regional, and internal (Maviza, 2020). Different push and pull factors have been identified to explain why people migrate. International migration has been driven by (1) the ever-widening global inequalities between countries and regions in the Global North and the Global South (Dávalos, 2012), (2) socioeconomic and political crises in most countries in the Global South, and (3) the perpetual demands for unskilled labour in the Global North (Adepoju, 2000; Flahaux & De Haas, 2016). Regional or South–South migration has mainly been fuelled by proximity, networks, and regional and bilateral agreements (Ratha & Shaw, 2007; Moyo, 2020a). In internal migration, mobility is spurred by the urban-biased colonial legacies of development in the Global South, which were maintained in the post-colonial era (Kurekova, 2011). These urban biases in development produced skewed spatial development between the rural and urban areas (Kurekova, 2011; Takyi, 2011). These imbalances continue to push people from rural areas to urban centres in search of better economic opportunities (Potts, 2010; Makombe, 2013; Bello-Bravo, 2015).

Although this has been the general trend, fragmented anecdotal evidence has shown that climate change has also become a notable driver of migration globally (Klepp, 2017; Borderon et al., 2019; Schraven et al., 2019). In the past decade, there has been a notable increase in scholarship assessing the impacts of climate change on human migration (Hoffmann et al., 2021). Although this has been the case, Hoffman et al. (2021) posit that findings from these studies vary significantly, which makes it difficult to conclude when climate-induced migration occurs. Notwithstanding, significant scholarship has indicated that the extreme events induced by climate change – droughts, floods, erratic rains, and high temperatures – and their effects on productivity across sectors are increasingly recognised as probable push factors in human mobility (Raleigh & Jordan, 2010; Schraven et al., 2019; Marotzke et al., 2020). This migration has been code-named climate migration to differentiate it from the other forms of mobility (Wyman, 2013; Bettini, 2014; Boas et al., 2019).

In essence, due to the adverse effects of climate change, migration has been adopted as a coping strategy. This is especially true for most African countries, where most economies are sustained by climate-sensitive sectors (Connolly-Boutin & Smit, 2016; Oduniyi, 2018). Therefore, with climate change, productivity in such sectors as agriculture may be negatively affected, leading to high levels of food insecurity and adverse effects on income and employment opportunities for some (Klepp, 2017). Thus, its impacts on agriculture the availability of water and energy make it a pressure point that pushes people to migrate to cities or other countries as a coping strategy (Schraven et al., 2019). Klepp (2017) proffered that experts on climate change expect an increased frequency of extreme weather events due to an increasingly warming world. This is expected to aggravate existing vulnerabilities related to food, water, energy, and health, among others, especially in the Global South. In Southern Africa, for example, climate change is expected to lead to a decrease in the annual mean rainfall and an increase in the incidence of erratic rainfall (Oduniyi, 2018; Maviza & Ahmed, 2021).

Given that most households rely on rain-fed agriculture for subsistence, its compromised productivity due to climate change and variation forces many to diversify their livelihood portfolios (Oduniyi, 2018). Significant scholarship shows that, as a result of the preceding, there is evidence of intensified circular and seasonal migration as a coping and adaptation strategy in the region (Oduniyi, 2018; Borderon et al., 2019). As this happens, the disproportionate burden of responsibility between men and women concerning households' daily sustenance exacerbates the gendered effects and impacts of climate change. As alluded to earlier, this may be one of the reasons to explain the increase in the feminisation of migration, as women may adopt mobility as a livelihood diversification strategy to ensure their children are supported.

In some instances, permanent migration has also emerged as a response from the economically active family members who leave in search of better economic opportunities (Oduniyi, 2018; Chidakwa et al., 2020). Notwithstanding, the biggest challenge in the African context has been ascertaining the actual cause of migration, given that migration has long been a significant characteristic of the Southern African Region. Therefore, although there are notable merits in the arguments that pin climate change as a driver of migration, there remains a need for robust research to ascertain the causal effects between the two, especially in Sub-Saharan Africa. However, the question that lingers beyond this is whether these mobilities foster social equity, especially for the families left behind. Migration typically divides families (Murray, 1981; Silver, 2006), and the consequences in some cases have been detrimental. Therefore, considering the climate change discourse and the need to foster social equity, it becomes critical to assess whether migration offers a positive pathway in the fight against climate change.

Migration as Means of Resilience in the Face of Livelihood Risks

Considering the increasing effects of climate change that threaten livelihoods, there is a need for affected communities to build resilience. Some countries in Southern Africa, such as Mozambique, were hit by multiple cyclones (i.e., Idai and Kenneth) in 2019 after experiencing a prolonged drought that reduced agricultural production by half (Mbiyozo, 2020). These two devastating storms displaced 685,000 people and had an impact that reached other countries that, included Comoros, Madagascar, Malawi, and Zimbabwe (Mbiyozo, 2020). Without alternative ways of building resilience, most affected communities rely on humanitarian aid, which does little to provide a long-term solution. Migration allows families to diversify their livelihoods and self-insure against livelihood risks such as those caused by climate-induced natural disasters (Nzima et al., 2016). Household members who may already be migrants working in other countries or within the country in unaffected cities can send remittances that can be used to rebuild after a devastating storm. In addition, in-kind remittances such as food items can also provide relief when communities face catastrophes such as floods. Migration can also be a household decision that allows communities to be disaster prepared by using remittances earned in better economies to build structures that are strong and resistant to water damage. Most communities affected by devastating storms have homes built using mud instead of concrete and cement bricks. Migrant remittances can be used to build more resilient homes that will withstand the storms.

Mbiyozo (2020) draws us to the lack of information as an important barrier to disaster preparedness. According to Mbiyozo (2020), many victims of cyclone Idai were not aware of climate change and the damage that extreme weather events can cause; as such, there was reluctance to heed warnings to move until severe damage in crops, homes and land had been suffered. We argue that migration can play an important role in bringing awareness to remote areas that often suffer from these climate-induced natural disasters. As people migrate, they gain knowledge and are exposed to information and skills that can be transferred to their origin villages. Migrants can bring awareness to their origin villages about possible climate-induced disasters on time to enhance preparedness and resilience. In addition, migrants can bring new skills and knowledge about new technologies that can improve agricultural production and through remittances they can finance these. These would undoubtedly strengthen resilience in the face of livelihood risks caused by climate change and in turn ensure that there is social equity in climate change response strategies.

Conclusion

Southern Africa is clearly among the sub-regions that are highly susceptible to climate-induced disasters such as prolonged droughts, rising temperature, and tropical cyclones and storms. In addition, the region has a widely documented history of migration, and considerable research continues to be done on migration. However, there is limited research that focuses on climate change-induced migration in the region. This can partly be attributed to the complexity of factors surrounding this link and identifying migration push factors related to climate change as these can be hidden in broader economic issues such as joblessness and structural inequalities. However, there is an urgent need to research on how the increasing effects of climate change influence migration patterns. There is also a need to trace the possible contributions made by climate change towards the changing gendered patterns of migration manifesting in the feminisation of migration. It is highly likely that as agro-based livelihoods are destroyed by droughts, water scarcity and flooding, women who have traditionally remained behind when men migrated may have joined to build resilience and ensure social equity in climate change response strategies. These women are engaged in cross-border trading, while some seek seasonal jobs in commercial farms in neighbouring countries. The incomes earned by women could be what is used to provide relief for families in poor communities that experience the harsh effects of extreme weather conditions. In dealing with the problem of climate change, it is crucial to ensure that interventions ensure social equity so that no one is left behind. This is especially important given the increasing number of female migrants in Southern Africa possibly owing to climate change that has affected agro-based livelihoods. In the interest of social equity, there is an urgent need to explore how remittances earned by migrants contribute to building resilience and how these can be directed towards investing in smart technologies that can enhance food production amid climate-induced natural disasters.

Notes

1 https://www.sadc.int/news-events/news/sadc-urges-support-rural-women-world-commemorates-international-womens-day-2018/.
2 https://www.sadc.int/news-events/news/sadc-urges-support-rural-women-world-commemorates-international-womens-day-2018/.

References

Abu, M., Codjoe, S. N. A., & Sward, J. (2014). Climate change and internal migration intentions in the forest-savannah transition zone of Ghana. *Population and Environment*, *35*(4), 341–364.

Adepoju, A. (2000). Issues and recent trends in international migration in Sub-Saharan Africa. *International Social Science Journal*, *52*(165), 383–394.

Adisa, T. A., Abdulraheem, I., & Isiaka, S. B. (2019). Patriarchal hegemony. *Gender in Management: An International Journal*, 34(1), 19–33.

Allwood, G. (2020). Mainstreaming gender and climate change to achieve a just transition to a climate neutral Europe. *Journal of Common Market Studies*, *58*(S1), 173–186.

Baldassar, L., & Merla, L. (Eds.). (2014). *Transnational families, migration and the circulation of care: Understanding mobility and absence in family life*. Routledge.

Bello-Bravo, J. (2015). Rural-urban migration: A path for empowering women through entrepreneurial activities in West Africa. *Journal of Global Entrepreneurship Research*, *5*(1), 9.

Bettini, G. (2014). Climate migration as an adaption strategy: De-securitizing climate-induced migration or making the unruly governable? *Critical Studies on Security*, *2*(2), 180–195.

Boas, I., Farbotko, C., Adams, H., Sterly, H., Bush, S., Van der Geest, K., Wiegel, H., Ashraf, H., Baldwin, A., & Bettini, G. (2019). Climate migration myths. *Nature Climate Change*, *9*(12), 901–903.

Borderon, M., Sakdapolrak, P., Muttarak, R., Kebede, E., Pagogna, R., & Sporer, E. (2019). Migration influenced by environmental change in Africa. *Demographic Research*, *41*, 491–544.

Brown, M. E., & Funk, C. C. (2008). Food security under climate change. *Science*, *319*(5863), 580–581.

Camill, P. (2010). Global change. *Nature Education Knowledge*, *2*(1), 49.

Camlin, C. S., Snow, R. C., & Hosegood, V. (2014). Gendered patterns of migration in rural South Africa. *Population, Space and Place*, *20*(6), 528–551.

Carling, J. (2005). Gender dimensions of international migration. *Global Migration Perspectives*, *35*, 1–26.

Chang, A. Y., Fuller, D. O., Carrasquillo, O., & Beier, J. C. (2014). Social justice, climate change, and dengue. *Health and Human Rights Journal*, 16, 93–104.

Chidakwa, P., Mabhena, C., Mucherera, B., Chikuni, J., & Mudavanhu, C. (2020). Women's vulnerability to climate change: Gender-skewed implications on agro-based livelihoods in rural Zvishavane, Zimbabwe. *Indian Journal of Gender Studies*, *27*(2), 259–281.

Chingarande, D., Huyer, S., Lanzarini, S., Makokha, J. N., Masiko, W., Mungai, C.,… & Waroga, V. (2020). *Background paper on mainstreaming gender into National Adaptation Planning and implementation in Sub-Saharan Africa*. CCAFS Working Paper.

Connolly-Boutin, L., & Smit, B. (2016). Climate change, food security, and livelihoods in sub Saharan Africa. *Regional Environmental Change*, *16*(2), 385–399.

Dankelman, I. (2010). *Gender and climate change: An introduction.* Routledge.

Dávalos, C. L. (2012). *Exploring transnational families among Ecuadorian migrant workers in Spain: The case of cleaners in Madrid.* Queen Mary University of London. http://qmro.qmul.ac.uk/xmlui/handle/123456789/8448.

Delius, P. (2017). The history of migrant labor in South Africa (1800–2014). *Oxford Research Encyclopedia of African History.* https://doi.org/10.1093/acrefore/9780190277734.013.93

Donaldson, M., & Howson, R. (2009). Men, migration and hegemonic masculinity. *Migrant Men: Critical Studies of Masculinities and the Migration Experience, 20,* 210.

Dube, T., Moyo, P., Ncube, M., & Nyathi, D. (2016). The impact of climate change on agro ecological based livelihoods in Africa: A review. *Journal of Sustainable Development, 9*(1), 256–267.

Flahaux, M.-L., & De Haas, H. (2016). African migration: Trends, patterns, drivers. *Comparative Migration Studies, 4*(1), 1.

Garutsa, T. C., Mubaya, C. P., & Zhou, L. (2018). Gendered differentials in climate change adaptation amongst the Shona ethnic group in Marondera Rural District, Zimbabwe: A social inclusions lens. *AAS Open Research, 1*(14), 14.

Gouws, A. (2010). The feminisation of migration. *Africa Insight, 40*(1), 169–180.

Helman, R., & Ratele, K. (2016). Everyday (in) equality at home: Complex constructions of gender in South African families. *Global Health Action, 9*(1), 31122.

Hoffmann, R., Šedová, B., & Vinke, K. (2021). Improving the evidence base: A methodological review of the quantitative climate migration literature. *Global Environmental Change, 71,* 102367.

Hunter, M. (2010). Beyond the male-migrant: South Africa's long history of health geography and the contemporary AIDS pandemic. *Health & Place, 16*(1), 25–33.

Jayachandran, S. (2015). The roots of gender inequality in developing countries. *Economics, 7*(1), 63–88.

Jury, M. R. (2013). Climate trends in southern Africa. *South African Journal of Science, 109*(1), 1–11.

Kendall, D. (2012). *Sociology in our times.* Cengage Learning.

Klepp, S. (2017). Climate change and migration. In S. Klepp (Ed.), *Oxford research encyclopedia of climate science.* Oxford University Press. https://doi.org/10.1093/acrefore/9780190228620.013.42.

Kurekova, L. (2011). Theories of migration: Conceptual review and empirical testing in the context of the EU East-West flows. *Interdisciplinary Conference on Migration, Economic Change, Social Challenge 4,* 6–9. University College London.

Kusangaya, S., Warburton, M. L., Van Garderen, E. A., & Jewitt, G. P. (2014). Impacts of climate change on water resources in southern Africa: A review. *Physics and Chemistry of the Earth, Parts A/B/C, 67,* 47–54.

Landau, L. (2011). *Contemporary migration to South Africa: A regional development issue.* World Bank Publications.

Lim Kam Sian, K. T. C., Wang, J., Ayugi, B. O., Nooni, I. K., & Ongoma, V. (2021). Multi decadal variability and future changes in precipitation over southern Africa. *Atmosphere, 12*(6), 742.

Mabhena, C. (2010). *'Visible hectares, vanishing livelihoods': A case of the Fast Track Land Reform and Resettlement Programme in southern Matabeleland-Zimbabwe.* (Doctoral dissertation, University of Fort Hare).

Makombe, E. K. (2013). A social history of town and country interactions: A study on the changing social life and practices of rural-urban migrants in colonial Harare and Goromonzi (1946 1979). (Doctoral dissertation, Wits University).

Mangleburg, T. F., Grewal, D., & Bristol, T. (1999). Family type, family authority relations, and adolescents' purchase influence. In E.J. Arnould and L.M. Scott (Eds.), *NA – Advances in Consumer Research* Volume 26, Provo, UT: Association for Consumer Research, 379–384.

Maphosa, F. (2007). Remittances and development: The impact of migration to South Africa on rural livelihoods in southern Zimbabwe. *Development Southern Africa, 24*(1), 123–136.

Maphosa, F. (2010). Transnationalism and undocumented migration between rural Zimbabwe and South Africa. In J. Crush, D. Tevera (Eds.), *Zimbabwe's Exodus: Crisis, Migration, Survival*. Ottawa and Kingston: IDRC and SAMP, 346–374).

Maphosa, F. (2012). Irregular migration and vulnerability to HIV&AIDS: Some observations from Zimbabwe. *Africa Development, 37*(2), 119–135.

Marotzke, J., Semmann, D., & Milinski, M. (2020). The economic interaction between climate change mitigation, climate migration and poverty. *Nature Climate Change, 10*(6), 518–525.

Masanja, G. F. (2012). The female face of migration in sub-Saharan Africa. *Huria: Journal of the Open University of Tanzania, 11*(1), 80–97.

Mashiri, L., & Mawire, P. R. (2013). Conceptualisation of gender based violence in Zimbabwe. *International Journal of Humanities and Social Science, 3*(15), 94–103.

Masinga, F. N., Maharaj, P., & Nzima, D. (2021). Adapting to changing climatic conditions: Perspectives and experiences of women in rural KwaZulu-Natal, South Africa. *Development in Practice, 31*(8), 1002–1013.

Mason, S. J. (2001). El Niño, climate change, and Southern African climate. *Environmetrics: The Official Journal of the International Environmetrics Society, 12*(4), 327–345.

Mastrorillo, M., Licker, R., Bohra-Mishra, P., Fagiolo, G., Estes, L. D., & Oppenheimer, M. (2016). The influence of climate variability on internal migration flows in South Africa. *Global Environmental Change, 39*, 155–169.

Matose, T., Maviza, G., & Nunu, W. (2022). Pervasive irregular migration and the vulnerabilities of irregular female migrants at Plumtree border post in Zimbabwe. *Journal of Migration and Health, 5*, 100091–100091.

Maunganidze, F. (2020). Dealing with gender-related challenges: A perspective of Zimbabwean women in the practice of law. *Cogent Business & Management, 7*(1), 1769806.

Maviza, A., & Ahmed, F. (2021). Climate change/variability and hydrological modelling studies in Zimbabwe: A review of progress and knowledge gaps. *SN Applied Sciences, 3*(5), 1–28.

Maviza, G. (2020). *Transnational migration and families—Continuities and changes along processes of sustained migration: A case of Tsholotsho in Matabeleland North Zimbabwe* [Doctor of Philosophy]. University of Witwatersrand.

Mbiyozo, A. N. (2020). Migration: A critical climate change resilience strategy. Policy Brief. Institute for Security Studies.

Mlambo, A. 2010. A History of Zimbabwean Migration to 1990. In J. Crush and D. Tevera (Eds.), *Zimbabwe's Exodus: Crisis, Migration, Survival*. Ottawa and Kingston: IDRC and SAMP, 52–76.

Moser, C. (2012). *Gender planning and development: Theory, practice and training*. Routledge.

Moyo, I. (2020a). On decolonising borders and regional integration in the Southern African Development Community (SADC) region. *Social Sciences, 9*(4), 32.

Moyo, I. (2020b). *Why South Africa's new plan to fortify its borders won't stop irregular migration*. The Conversation Africa. https://theconversation.com/why-south-africas-new plan-to-fortify-its-borders-wont-stop-irregular-migration-145072.

Moyo, K. (2017). *Zimbabweans in Johannesburg, South Africa: Space, movement and spatial identity*. University of Witwatersrand.

Murray, C. (1981). *Families divided: The impact of migrant labour in Lesotho* (Vol. 29). Cambridge University Press.

Musoni, F. (2020). *Border jumping and migration control in Southern Africa*. Indiana University Press.

Nagel, J. (2015). *Gender and climate change: Impacts, science, policy*. Routledge.

Ncube, G., Dube, N., & Sithole, M. (2014). Immigration policy reforms and pervasive illegal migration: A case of the illegal border jumpers (Beitbridge border post). *IOSR Journal of Humanities and Social Science*, *19*(5), 114–121.

Ndlovu, D. S., & Landau, L. B. (2020). The Zimbabwe–South Africa migration corridor. In T. Bastia and R. Skeldon (Eds.), *Routledge Handbook of Migration and Development*, London: Routledge, 474–478.

Nhamo, L., Matchaya, G., Mabhaudhi, T., Nhlengethwa, S., Nhemachena, C., & Mpandeli, S. (2019). Cereal production trends under climate change: Impacts and adaptation strategies in southern Africa. *Agriculture*, *9*(2), 30.

Nyamunda, T. (2014). Cross-border couriers as symbols of regional grievance?: The Malayitsha Remittance System in Matabeleland, Zimbabwe. *African Diaspora*, *7*(1), 38–62.

Nzima, D., Duma, V., & Moyo, P. (2016). Migrant remittances, livelihoods and investment: Evidence from Tsholotsho District in the Matabeleland North Province of Zimbabwe. *Migracijske i etničke teme*, *32*(1), 37–62.

Nzima, D., & Moyo, P. (2017). The new 'diaspora trap' framework: Explaining return migration from South Africa to Zimbabwe beyond the 'failure-success' framework. *Migration Letters*, *14*(3), 355–370.

Oduniyi, O. S. (2018). Implication of climate change on livelihood and adaptation of small and emerging maize farmers in the North West Province of South Africa. *Doctoral Thesis submitted at the University of South Africa*.

Patel, J. (2020). *Despite perceptions of gender equality, Africa's rural women bear brunt of economic exclusion*. Afrobarometer Dispatch No. 397. Cape Town: Institute for Justice and Reconciliation.

Pearse, R. (2017). Gender and climate change. *Wiley Interdisciplinary Reviews: Climate Change*, *8*(2), e451.

Pillinger, J. (2007). *The feminisation of migration*. Immigrant Council of Ireland.

Pinnawala, M. (2009). *Gender transformation and female migration*. (Doctoral dissertation, University of Peradeniya, Sri Lanka).

Posel, D. (2003). The collection of national household survey data in South Africa (1993–2001): Rendering labour migration invisible. *Development Southern Africa*, *20*(3), 361–368.

Posel, D., & Marx, C. (2013). Circular migration: A view from destination households in two urban informal settlements in South Africa. *The Journal of Development Studies*, *49*(6), 819–831.

Potts, D. (2010). Internal migration in Zimbabwe: The impact of livelihood destruction in rural and urban areas. In. In J. Crush, D. Tevera (Eds.), *Zimbabwe's Exodus Southern*

African Migration Project. Cape Town: International Development Research Centre, 79–109.

Potts, D., & Mutambirwa, C. (1990). Rural-urban linkages in contemporary Harare: Why migrants need their land. *Journal of Southern African Studies, 16*(4), 677–698.

Preston, I., Banks, N., Hargreaves, K., Kazmierczak, A., Lucas, K., Mayhe, R., Street, R. (2014). *Climate Change and Social Justice: An Evidence Review*. York: Joseph Rowntree Foundation

Radel, C., & Schmook, B. (2009). Migration and gender: The case of a farming ejido in Calakmul, Mexico. *Yearbook of the Association of Pacific Coast Geographers, 71*(1), 144–163.

Raleigh, C., & Jordan, L. (2010). Climate change and migration: Emerging patterns in the developing world. In R. Mearns, A. Norton (Eds.), *Social Dimensions of Climate Change: Equity and Vulnerability in a Warming World* , Washington DC: World Bank, pp. 103–131.Ratha, D., & Shaw, W. (2007). Causes of South-South migration and its socioeconomic effects. *Migration Information Source, 17*. http://www.migrationinformation.org/feature/display.cfm?ID=647

Schraven, B., Adaawen, S., Rademacher-Schulz, C., & Segadl, N. (2019). *Human mobility in the context of climate change in Sub-Saharan Africa: Trends and basic recommendations for development cooperation*. Bonn: German Development Institute.

Scott, P. (1954). Migrant labor in southern Rhodesia. *Geographical Review, 44*(1), 29–48.

Silver, A. (2006). *Families across borders: The effects of migration on family members remaining at home* (Doctoral Dissertation, The University of North Carolina at Chapel Hill).

Strobl, E., & Valfort, M. A. (2015). The effect of weather-induced internal migration on local labor markets. Evidence from Uganda. *The World Bank Economic Review, 29*(2), 385–412.

Takaindisa, J. (2020). *Transnational mothering, patterns and strategies of care-giving by Zimbabwean domestic workers in Botswana: A multi-sited approach* [Doctor of Philosophy]. University of the Witwatersrand.

Takyi, B. K. (2011). *Transformations in the African family: A note on migration, HIV/AIDS and family poverty alleviation efforts in Sub-Saharan Africa (SSA)*. In *United Nations Expert Group Meeting on Assessing Family Policies*, New York, pp. 1–3.

Thebe, P. (2019). Determinants of feminization of migration in Tsholotsho District of Zimbabwe. *Advances in Social Sciences Research Journal, 6*(10), 297–306.

Thebe, P., & Maviza, G. (2019). The effects of feminization of migration on family functions in Tsholotsho District, Zimbabwe. *Advances in Social Sciences Research Journal, 6*(5), 297–306

Thoithi, W., Blamey, R. C., & Reason, C. J. (2021). Dry spells, wet days, and their trends across Southern Africa during the summer rainy season. *Geophysical Research Letters, 48*(5), e2020GL091041.

Tshabalala, X. (2019). Hyenas of the Limpopo: "Illicit labour recruiting," assisted border crossings, and the social politics of movement across South Africa's border with Zimbabwe. *Journal of Borderlands Studies, 34*(3), 433–450.

Wentzel, M. (2003). *Historical and contemporary dimensions of migration between South Africa and its neighbouring countries*. Unpublished Paper Delivered at HSRC Migration Workshop, Pretoria, 17–20 March 2003.

Wyman, K. M. (2013). Responses to climate migration. *Harvard Environmental Law Review, 37*, 167.

Yabiku, S. T., Agadjanian, V., & Sevoyan, A. (2010). Husbands' labour migration and wives' autonomy, Mozambique 2000–2006. *Population Studies*, *64*(3), 293–306.

Yeates, N. (2004). Global care chains. *International Feminist Journal of Politics*, *6*(3), 369–391. https://doi.org/10.1080/1461674042000235573.

Zack, T., Matshaka, S., Moyo, K., & Vanyoro, K. P. (2019). *My way? The circumstances and intermediaries that influence the migration decision-making of female Zimbabwean domestic workers in Johannesburg*, Working Paper, 57, Migrating Out of Poverty, University of Sussex, Brighton.

Zinyama, L. M. (1990). International migrations to and from Zimbabwe and the influence of political changes on population movements, 1965–1987. *International Migration Review*, *24* (4), 748–767.

Index

Note: **Bold** page numbers refer to tables and *italic* page numbers refer to figures.

For Product Safety Concerns and Information please contact our EU
representative GPSR@taylorandfrancis.com
Taylor & Francis Verlag GmbH, Kaufingerstraße 24, 80331 München, Germany

www.ingramcontent.com/pod-product-compliance
Lightning Source LLC
Chambersburg PA
CBHW060250220326
41598CB00027B/4053

*9 7 8 1 0 3 2 5 0 1 6 1 1 *